SpringerBriefs in Mathematical Physics

Volume 22

SpringerBriefs are characterized in general by their size (50-125 pages) and fast production time (2-3 months compared to 6 months for a monograph).

Briefs are available in print but are intended as a primarily electronic publication to be included in Springer's e-book package.

Typical works might include:

- An extended survey of a field
- A link between new research papers published in journal articles
- A presentation of core concepts that doctoral students must understand in order to make independent contributions
- Lecture notes making a specialist topic accessible for non-specialist readers.

SpringerBriefs in Mathematical Physics showcase, in a compact format, topics of current relevance in the field of mathematical physics. Published titles will encompass all areas of theoretical and mathematical physics. This series is intended for mathematicians, physicists, and other scientists, as well as doctoral students in related areas.

More information about this series at http://www.springer.com/series/11953

Yuji Kodama

KP Solitons
and the Grassmannians

Combinatorics and Geometry
of Two-Dimensional Wave Patterns

 Springer

Yuji Kodama
The Ohio State University
Columbus, OH
USA

ISSN 2197-1757 ISSN 2197-1765 (electronic)
SpringerBriefs in Mathematical Physics
ISBN 978-981-10-4093-1 ISBN 978-981-10-4094-8 (eBook)
DOI 10.1007/978-981-10-4094-8

Library of Congress Control Number: 2017933856

Printed on acid-free paper

This Springer imprint is published by Springer Nature
The registered company is Springer Nature Singapore Pte Ltd.
The registered company address is: 152 Beach Road, #21-01/04 Gateway East, Singapore 189721, Singapore

To Mie, and our children, Lay and May

Preface

It is a well-known story that in August 1834, John Scott Russel observed a large solitary wave in a shallow water channel in Scotland. He notes in his first paper [109] on the subject that

> I was observing the motion of a boat which was rapidly drawn along a narrow channel by a pair of horses, when the boat suddenly stopped - not so the mass of water in the channel which it had put in motion; it accumulated round the prow of the vessel in a state of violent agitation, then suddenly leaving it behind, rolled forward with great velocity, assuming the form of a large solitary elevation, a rounded, smooth and well defined heap of water, which continued its course along the channel apparently without change of form or diminution of speed...

This solitary wave is now known as an example of a *soliton* and is described by a solution of the Korteweg-de Vries (KdV) equation [137]. The KdV equation describes one-dimensional wave propagation, such as beach waves parallel to the coastline or waves in narrow canal, and is obtained in the leading order approximation of an asymptotic perturbation theory under the assumptions of weak non-linearity (small amplitude) and weak dispersion (long waves). The KdV equation has rich mathematical structure, including the existence of N-soliton solutions and the Lax pair for the inverse scattering method, and it is a prototype equation of the $1 + 1$ dimensional integrable systems (see, e.g., [1, 2, 22, 48, 78, 89, 97, 98]).

In 1973, Kadomtsev and Petviashvili [61] proposed a $2 + 1$ dimensional dispersive wave equation to study the stability of one-soliton solution of the KdV equation under the influence of weak transversal perturbations in the y-direction. This equation is now referred to as the KP equation, and it is given by

$$\frac{\partial}{\partial x}\left(-4\frac{\partial u}{\partial t} + 6u\frac{\partial u}{\partial x} + \frac{\partial^3 u}{\partial x^3}\right) + 3\frac{\partial^2 u}{\partial y^2} = 0,$$

where $u = u(x, y, t)$ presents the wave amplitude at the point (x, y) in the xy-plane for fixed time t. The KP equation can be used to describe shallow water waves (see, e.g., [1, 70, 79]), and in particular, the equation provides an excellent model for the *resonant* interaction of those waves [74, 87, 88]. It turns out that the KP equation has a much richer structure than the KdV equation, and is considered to be the most fundamental integrable system in the sense that many known integrable systems can be derived as special reductions of the KP hierarchy, which is defined as the set of the KP equation together with its infinitely many symmetries (see, e.g., [2, 37, 63, 75, 98]). The KP equation has been recognized to be related to various areas of mathematics and physics, such as algebraic and enumerative geometry, representation theory, random matrix theory, and quantum field theory (there are numerous papers related to these topics; see, e.g., [5, 6, 8, 12, 23, 46, 54, 60, 64, 67, 76, 95, 96, 117, 129, 134] and the references therein).

One of the main breakthroughs in the KP theory was given by Sato in [111–114], who realized that solutions of the KP equation could be written in terms of points of an infinite-dimensional Grassmannian. This book deals with a real, finite-dimensional version of the Sato theory. In particular, we are interested in solutions that are *regular* in the entire *xy*-plane, where they are localized along certain rays. Such solutions are called *line-soliton solutions* or *KP solitons* in this book, and they can be constructed from points of the real finite-dimensional Grassmannian.

Because of the nonlinearity in the KP equation, the solutions form very complex web-like patterns in the *xy*-plane which are generated by resonant interactions among several obliquely propagating line solitons. The set of figures below illustrates an example of such patterns. Each figure shows the contour plot of the solution at a fixed time *t* in the *xy*-plane with *x* in the horizontal and *y* in the vertical directions.

The main aim of this book is to give a geometric and combinatorial classification of the patterns generated by the line-soliton solutions as in the figures. There are mainly three parts: In Chaps. 1–3, I provide a brief introduction of the Sato theory of the KP equation and the KP solitons, which is the main subject of this book. Chapters 4 and 5 give an invitation to the *totally nonnegative* Grassmannians and present their parameterizations, which provides the mathematical foundation of the KP solitons. Then, Chaps. 6–8 present a classification theorem of the KP solitons and describe the structure of the spatial patterns, referred to as *soliton graphs*, generated by the KP solitons.

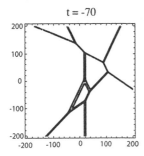

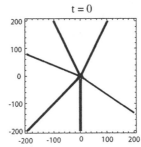

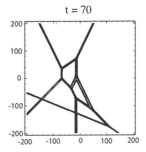

John Scott Russel continues on in his book [110] to say that

This is a most beautiful and extraordinary phenomenon: the first day I saw it was the happiest day of my life. Nobody has ever had the good fortune to see it before or, at all events, to know what it meant. It is now known as the solitary wave of translation. No one before had fancied a solitary wave as a possible thing.

I hope the present book is successful in convincing you (the readers) that "*this (two-dimensional wave pattern generated by the KP equation) is a most beautiful and extraordinary phenomena*" of two-dimensional nonlinear wave dynamics, and have no doubt that observing it, for example at a beach or even in a numerical simulation of shallow water waves, will be one of the *happiest* moments of your life.

For more than 10 years, I have been working on the subject related to this book with several people. I am most grateful to my research collaborators Sarbarish Chakravarty, Lauren Williams, Ken-Ichi Maruno, Harry Yeh, Chiu-Yen Kao, Masayuki Oikawa, Hidekazu Tsuji, Gino Biondini, and my former Ph.D. students, Yuhan Jia, and Jihui Huang.

Most of the materials in this book are based on several series of lectures. The first two series were given at the Chinese Academy of Science in Beijing, June of 2008 and July of 2009. I would like to thank Qing-Ping Liu, Xing-Biao Hu, and Ke Wu for the invitation, and Tian Kai for taking lecture notes.

In the winter quarter of 2010 at the Ohio State University, I delivered a graduate class titled "*Nonlinear Waves,*" which gives an overview of the background information for Chaps. 1 and 2 of this book. I would like to thank the students in the class for their comments and discussions. In particular, I am grateful to Chuanzhong Li, a visiting student from China, for his many useful suggestions.

During the second half of the year of 2012, I visited several institutes and Universities to give a series of lectures on the subjects related to this book. First, in June, I was invited to give a series of lectures titled "*Mathematical foundation of integrable systems and their applications*" at the University of Roma Tre, and would like to thank Decio Levi for the invitation and discussions. Then, in July, I was at the Institute of Mathematics, Academia Sinica in Taipei, to present "*Real Grassmannian and KP solitons.*" I would like to thank Jyh-Hao Lee and Derchyi Wu for the invitation and their kind hospitality during my stay in Taipei. In September, I delivered a series of lectures titled "*Real Grassmann varieties and their applications to integrable systems*" to the graduate students of Kyushu University in Japan, and would like to thank Hiroaki Hiraoka for the invitation and his kind hospitality. Then in October, I gave a similar lecture series at Nagoya University and would like to thank Masashi Hamanaka for his kind hospitality during my stay in Nagoya.

I was then selected to be a principal speaker at the NSF-CBMS conference at UTPA in May, 2013, and delivered 10 lectures on "*Solitons in two-dimensional water waves and applications to tsunami.*" I would very much like to thank Ken-Ichi Maruno and Virgil Pierce for organizing such an excellent conference.

I am very grateful to Division of Mathematical Sciences of National Science Foundation for the generous and encouraging supports for my research on the KP solitons and related subjects: NSF grants, DMS-0806219, DMS-1108813, and DMS-1410267.

My thanks also go to Prof. Atsuo Kuniba for inviting me to write this monograph, and Mr. Masayuki Nakamura for his kind assistance during the preparation of this book. Various comments and suggestions of anonymous referees are greatly appreciated. I would also like to thank my recent collaborator Rachel Karpman for reading the book and making several useful comments.

Special thanks go to my daughters, Lay and May, for their hard work on editing their daddy's English.

Columbus, OH, USA   Yuji Kodama

Contents

Chapter 1
Introduction to KP Theory and KP Solitons

Abstract We begin with a study of the Burgers equation, which is the simplest equation combining both nonlinear propagation effects and diffusive effects, and can be used to describe a weak shock phenomena in gas dynamics (see e.g. [131]). The Burgers equation can be linearized by a nonlinear transformation, known as the *Cole-Hopf transformation*. The linearization then shows that the Burger equation has an infinite number of symmetries, and the set of those symmetries defines the Burgers *hierarchy*. The linearization enables us to construct several exact solutions such as multi-shock solutions. It turns out that the set of those exact solutions forms a subclass of the solutions of the KP equation, and multi-shock solutions give examples of the resonant interactions in the KP solutions. We then extend the Cole-Hopf transformation to construct a multi-component Burgers hierarchy, and introduce the τ-function, which generates a large class of exact solutions of the KP equation, referred to as *KP solitons*. Based on the study of this multi-component Burgers hierarchy, we explain its connection to the *Sato theory*, which provides a mathematical foundation of the KP hierarchy in terms of an infinite dimensional Grassmann variety called the *Sato Grassmannian* [112, 113]. In this book, we consider a *finite* dimensional version of the Sato Grassmannian and construct each KP soliton from a point of this Grassmannian.

1.1 The Burgers Hierarchy

In this section, we demonstrate that some particular solutions of the KP equation can be obtained by solving the *Burgers hierarchy*. The Burgers equation describes a weak shock solution in a compressible fluid with a small diffusion term, and the Burgers hierarchy consists of the symmetries of the Burgers equation. It is quite remarkable that the Burgers hierarchy is intimately related to the KP equation. This fact explains the resonant phenomena that appears in the solutions of the KP equation as we will explain below.

© The Author(s) 2017
Y. Kodama, *KP Solitons and the Grassmannians*,
SpringerBriefs in Mathematical Physics 22, DOI 10.1007/978-981-10-4094-8_1

1

The Burgers equation is the following nonlinear partial differential equation for a function $w(x, t_2)$,

$$\frac{\partial w}{\partial t_2} = \frac{\partial^2 w}{\partial x^2} + 2w\frac{\partial w}{\partial x} = \frac{\partial}{\partial x}\left(\frac{\partial w}{\partial x} + w^2\right).$$

The Cole-Hoph transformation, which linearizes the Burgers equation, is given by (see, for example, Chap. 4 in [131]),

$$w = \frac{\partial}{\partial x}\ln f = \frac{f_x}{f}.$$

The transformation then leads to

$$\frac{\partial w}{\partial t_2} = \frac{\partial}{\partial t_2}\left(\frac{f_x}{f}\right) = \frac{\partial}{\partial x}\left(\frac{f_{t_2}}{f}\right) = \frac{\partial}{\partial x}\left(\frac{\partial}{\partial x}\left(\frac{f_x}{f}\right) + \left(\frac{f_x}{f}\right)^2\right) = \frac{\partial}{\partial x}\left(\frac{f_{xx}}{f}\right).$$

Integrating over x for the third and the last terms, we have

$$f_{t_2} = f_{xx} + c(t_2)f,$$

where $c(t_2)$ is an arbitrary function of t_2. This can be absorbed into f by the scaling $f\,e^{\int^{t_2} c(s)ds}$, so that we set $c = 0$. That is, the function f satisfies the *linear* diffusion equation. Thus we have the following:

Proposition 1.1 *The Burgers equation can be linearized by the Cole-Hopf transformation up to a scaling, and we have the compatibility relation,*

$$\begin{cases} f_x = wf \\ f_{t_2} = f_{xx} \end{cases} \quad\Longleftrightarrow\quad w_{t_2} = w_{xx} + 2ww_x.$$

We now define the Burgers *hierarchy*. Let $\{t_n : n = 1, 2, \ldots\}$ be multi-time variables defined by

$$\frac{\partial f}{\partial t_n} = \frac{\partial^n f}{\partial x^n} =: \partial_x^n f \quad \text{with} \quad x = t_1.$$

Notice that these equations are mutually compatible, i.e. $f_{t_i t_j} = f_{t_j t_i}$ for any i, j.

Then the Burgers hierarchy is defined by the compatibility between $f_x = wf$ and $f_{t_n} = \partial_x^n f$, and we have (Problem 1.1)

$$\frac{\partial w}{\partial t_n} = \frac{\partial}{\partial x}\left(\frac{\partial}{\partial x} + w\right)^n 1, \quad \text{for} \quad n = 1, 2, 3, \ldots. \tag{1.1}$$

If we set $y = t_2$ and $t = t_3$, then we have the following theorem.

Theorem 1.1 *Any solution of the Burgers hierarchy is also a solution of the KP equation, i.e.*

$$\begin{cases} f_x = wf \\ f_y = f_{xx} \\ f_t = f_{xxx} \end{cases} \implies u = 2w_x = 2(\ln f)_{xx} \quad \text{satisfies the KP equation}$$

Proof First note that the first two members of the Burgers hierarchy are given by

$$w_y = (w_x + w^2)_x \,,$$
$$w_t = (w_{xx} + 3ww_x + w^3)_x \,.$$

Then taking the derivatives, we have

$$w_{yy} = (w_{xxx} + 2w_x^2 + 4ww_{xx} + 4w^2 w_x)_x$$
$$w_{xt} = (w_{xxx} + 3w_x^2 + 3ww_{xx} + 3w^2 w_x)_x$$

Calculating $3w_{yy} - 4w_{xt}$, we have the potential KP equation for w, i.e.

$$3w_{yy} - 4w_{xt} = (-w_{xxx} - 6w_x^2)_x,$$

which gives the KP equation for $u = 2w_x = 2(\ln f)_{xx}$,

$$(-4u_t + u_{xxx} + 6uu_x)_x + 3u_{yy} = 0. \tag{1.2}$$

Thus, if w is the solution of the Burgers hierarchy with $t_2 = y$ and $t_3 = t$, the function $u = 2w_x$ is a solution of the KP equation.   □

One should note that in the y-direction the solution has a dissipative behavior, and this leads to a confluence of two shocks given by the Burgers equation (see Sect. 4.7 in [131], also see Sect. 1.1.2). The dissipative property of the Burgers equation in the y-direction should not be confused with the *dispersive* nature of the KP equation in the physical time t-direction.

Since $f(x, y, t)$ satisfies the *linear* equations in y and t, the general solution can be written in terms of the Fourier transform,

$$f(x, y, t) = \int_C e^{kx + k^2 y + k^3 t} \, d\mu(k),$$

where C is a curve with the parameter $k \in \mathbb{C}$ and $d\mu(k)$ is an appropriate measure on that curve. For a special solution of the Burgers equation, we consider a finite dimensional solution (finite Fourier transform) with the measure $d\mu(k) = \sum_{j=1}^M \rho_j \delta(k - \kappa_j) \, dk$ for real parameters $\rho_j, \kappa_j \in \mathbb{R}$. Then the function f is given by

$$f(x, y, t) = \sum_{j=1}^{M} \rho_j \, E_j(x, y, t) \quad \text{with} \quad E_j := e^{\kappa_j x + \kappa_j^2 y + \kappa_j^3 t}. \tag{1.3}$$

Note that if $M = 1$, the solution becomes trivial, $u = 0$. The solution generated by (1.3) is a multi-shock solution of the Burgers equation. In the following sections, we explain some of the properties of those solutions and discuss their connection to the KP equation.

1.1.1 Shock Solution and Line-Soliton Solution

Let us express a particular solution of the Burgers equation and its connection to one-soliton solution of the KP equation. Here we also introduce some notations for the soliton solution.

We take the formula (1.3) with $M = 2$, where we need a condition $\rho_1 \rho_2 > 0$ for a *regular* solution. Without loss of generality, we assume both $\rho_i > 0$, and write

$$\begin{aligned} f &= \rho_1 E_1 + \rho_2 E_2 = e^{\theta_1} + e^{\theta_2} \\ &= e^{\frac{1}{2}(\theta_1 + \theta_2)} \left(e^{\frac{1}{2}(\theta_1 - \theta_2)} + e^{-\frac{1}{2}(\theta_1 - \theta_2)} \right) \\ &= 2e^{\frac{1}{2}(\theta_1 + \theta_2)} \cosh \tfrac{1}{2}(\theta_1 - \theta_2) \end{aligned}$$

where θ_i's are given by $\theta_i = \kappa_i x + \kappa_i^2 y + \kappa_i^3 t + \ln \rho_i$.

Then we have

$$w = (\ln f)_x = \tfrac{1}{2}(\kappa_1 + \kappa_2) + \tfrac{1}{2}(\kappa_1 - \kappa_2) \tanh \tfrac{1}{2}(\theta_1 - \theta_2),$$

which represents a *shock* solution of the Burgers equation. When $\kappa_1 < \kappa_2$, the solution behaves

$$w(x, y, t) \rightarrow \begin{cases} \kappa_1, \text{ for } x \ll 0, \\ \kappa_2, \text{ for } x \gg 0. \end{cases}$$

That is, the solution is a monotone increasing function of x for each fixed (y, t). This monotonicity is true for any (regular) soliton solution discussed in this book.

The shock solution gives a solution of the KP equation through $u = 2w_x = 2(\ln f)_{xx}$,

$$u = \tfrac{1}{2}(\kappa_1 - \kappa_2)^2 \operatorname{sech}^2 \tfrac{1}{2}(\theta_1 - \theta_2) \tag{1.4}$$

Figure 1.1 illustrates three-dimensional figure of u and the contour plot. The line of the wave crest (or peak) is given by $\theta_1 = \theta_2$, i.e.

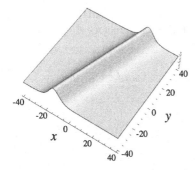 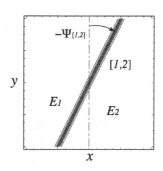

Fig. 1.1 One line-soliton solution labeled as [1, 2]-soliton. E_i represents the *dominant* exponential term in that region

$$x + (\kappa_1 + \kappa_2)y + (\kappa_1^2 + \kappa_2^2 + \kappa_1\kappa_2)t + \frac{1}{\kappa_1 - \kappa_2} \ln \frac{\rho_1}{\rho_2} = 0$$

Thus the soliton solution is localized along the line $\theta_1 = \theta_2$; hence we call it *line-soliton* solution. We emphasize here that the line-soliton appears at the boundary of two regions where either $E_1 = e^{\theta_1}$ or $E_2 = e^{\theta_2}$ is the dominant exponential term in the function f, and because of this we refer to this solution as [1, 2]-*soliton solution* or simply [1, 2]-*soliton*.

In this book, we construct more general line-soliton solutions, which consist of several line-solitons. Each of these line-solitons can be identified as [i, j]-soliton for some parameters $\kappa_i < \kappa_j$. The [i, j]-soliton solution has the same (local) structure as the one-soliton solution, and is described in the form,

$$u = A_{[i,j]} \operatorname{sech}^2 \frac{1}{2} \left(\mathbf{K}_{[i,j]} \cdot \mathbf{x} - \Omega_{[i,j]}t + \Theta^0_{[i,j]} \right)$$

with some constant $\Theta^0_{[i,j]}$. The amplitude $A_{[i,j]}$, the wave-vector $\mathbf{K}_{[i,j]}$ and the frequency $\Omega_{[i,j]}$ are defined by

$$\begin{cases} A_{[i,j]} = \frac{1}{2}(\kappa_j - \kappa_i)^2 \\ \mathbf{K}_{[i,j]} = \left(\kappa_j - \kappa_i, \kappa_j^2 - \kappa_i^2\right) = (\kappa_j - \kappa_i)\left(1, \kappa_i + \kappa_j\right), \\ \Omega_{[i,j]} = -(\kappa_j^3 - \kappa_i^3) = -(\kappa_j - \kappa_i)(\kappa_i^2 + \kappa_i\kappa_j + \kappa_j^2). \end{cases}$$

The direction of the wave-vector $\mathbf{K}_{[i,j]} = (K^x_{[i,j]}, K^y_{[i,j]})$ is measured in the counter-clockwise direction from the y-axis, and it is given by

$$\frac{K^y_{[i,j]}}{K^x_{[i,j]}} = \tan \Psi_{[i,j]} = \kappa_i + \kappa_j,$$

that is, $\Psi_{[i,j]}$ gives the angle between the line $\mathbf{K}_{[i,j]} \cdot \mathbf{x} = const$ and the y-axis. Then one line-soliton can be written in the form with three parameters $A_{[i,j]}$, $\Psi_{[i,j]}$ and $x^0_{[i,j]}$,

$$u = A_{[i,j]} \operatorname{sech}^2 \sqrt{\frac{A_{[i,j]}}{2}} \left(x + \tan \Psi_{[i,j]} y + C_{[i,j]} t - x^0_{[i,j]} \right), \qquad (1.5)$$

with $C_{[i,j]} = \kappa_i^2 + \kappa_i \kappa_j + \kappa_j^2 = \frac{1}{2} A_{[i,j]} + \frac{3}{4} \tan^2 \Psi_{[i,j]}$.

One should note here that the wave vector $\mathbf{K}_{[i,j]}$ and the frequency $\Omega_{[i,j]}$ satisfy the soliton-dispersion relation,

$$-4\Omega_{[i,j]} K^x_{[i,j]} = (K^x_{[i,j]})^4 + 3(K^y_{[i,j]})^2, \qquad (1.6)$$

which gives the relation in the parameters $(K^x_{[i,j]}, K^y_{[i,j]}, \Omega_{[i,j]})$ for the wave form $u = \exp(K^x_{[i,j]} x + K^y_{[i,j]} y - \Omega_{[i,j]} t)$ of the linearized KP equation $(-4u_t + u_{xxx})_x + 3u_{yy} = 0$. The soliton velocity $\mathbf{V}_{[i,j]}$ is along the direction of the wave-vector $\mathbf{K}_{[i,j]}$ and is defined by $\mathbf{K}_{[i,j]} \cdot \mathbf{V}_{[i,j]} = \Omega_{[i,j]}$, which yields

$$\mathbf{V}_{[i,j]} = \frac{\Omega_{[i,j]}}{|\mathbf{K}_{[i,j]}|^2} \mathbf{K}_{[i,j]} = -\frac{\kappa_i^2 + \kappa_i \kappa_j + \kappa_j^2}{1 + (\kappa_i + \kappa_j)^2} (1, \kappa_i + \kappa_j).$$

Note in particular that since $C_{[i,j]} = \kappa_i^2 + \kappa_i \kappa_j + \kappa_j^2 > 0$, the x-component of the soliton velocity is *always* negative, i.e., any soliton propagates in the negative x-direction. On the other hand, one should note that any small (linear) perturbation propagates in the positive x-direction, i.e., the x-component of the group velocity is always positive. This can be seen from the dispersion relation of the *linearized* KP equation for a plane wave $\phi = \exp(i\mathbf{k} \cdot \mathbf{x} - i\omega t)$ with the wave-vector $\mathbf{k} = (\kappa_x, \kappa_y)$ and the frequency ω,

$$\omega = \frac{1}{4} k_x^3 - \frac{3}{4} \frac{k_y^2}{k_x},$$

from which the group velocity of the wave is given by

$$\mathbf{v} = \nabla \omega = \left(\frac{\partial \omega}{\partial k_x}, \frac{\partial \omega}{\partial k_y} \right) = \left(\frac{3}{4} \left(k_x^2 + \frac{k_y^2}{k_x^2} \right), -\frac{3}{2} \frac{k_y}{k_x} \right).$$

This is similar to the case of the KdV equation ($k_y = 0$), and we expect that asymptotically, the soliton separates from small radiations. For certain initial data, the separation between solitons and radiations in the KP solution has been demonstrated numerically in [62]. However, at the present time, there is no rigorous analysis of the KP equation for general initial data. The reader may refer to the following references [19, 92, 93, 121, 130] and the references therein.

Remark 1.1 In the formulas (1.1), i.e. $f_{t_n} = \partial_x^n f$, if we include the higher times t_n in the exponential functions, i.e.

$$E_j(\mathbf{t}) = e^{\theta_j(\mathbf{t})}, \quad \text{with} \quad \theta_j(\mathbf{t}) := \sum_{n=1}^{\infty} \kappa_j^n t_n, \tag{1.7}$$

then the f-function gives a solution to the Burgers hierarchy so as to the KP hierarchy.

1.1.2 Confluence of Shocks: Resonant Interaction of Line-Solitons

It is well-known that when a shock overtakes another shock, they merge into a single shock of increased amplitude (see e.g. [131]). This phenomena can be observed in the y-evolution of a solution of the Burgers equation. Let us consider the solution $w = (\ln f)_x$ with $M = 3$ in (1.3),

$$f = \rho_1 E_1 + \rho_2 E_2 + \rho_3 E_3,$$

with some positive constants ρ_i for $i = 1, 2, 3$. As in the previous example, it is also possible here to determine the dominant exponentials and analyze the structure of the solution in the xy-plane. Let us consider the function f along the line $x + cy = 0$ with $c = \tan \Psi$ where Ψ is the angle measured counterclockwise from the y-axis (see Fig. 1.1). Then along $x + cy = 0$, we have the exponential function $E_j = \exp[\eta_j(c)y + \kappa_j^3 t]$ with

$$\eta_j(c) = \kappa_j(\kappa_j - c). \tag{1.8}$$

It is then seen from Fig. 1.2 that for $y \gg 0$ and a fixed t, the exponential term E_1 dominates when c is large positive number ($\Psi \approx \frac{\pi}{2}$, i.e. $x \to -\infty$). Decreasing the value of c (rotating the line clockwise), the dominant term changes to E_3. Thus we have

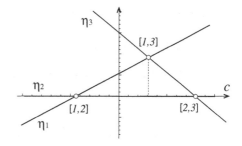

Fig. 1.2 The graphs of $\eta_j(c) = \kappa_j(\kappa_j - c)$. Each $[i, j]$ represents the exchange of the dominance order between η_i and η_j. The κ-parameters are given by $(\kappa_1, \kappa_2, \kappa_3) = (-0.3, 0, 0.5)$

$$w = \partial_x \ln f \longrightarrow \begin{cases} \kappa_1 & \text{as} \quad x \to -\infty, \\ \kappa_3 & \text{as} \quad x \to \infty. \end{cases}$$

The transition of the dominant exponentials $E_1 \to E_3$ is characterized by the condition $\eta_1 = \eta_3$, which corresponds to the direction parameter value $c = \tan \Psi_{[1,3]} = \kappa_1 + \kappa_3$. In the neighborhood of this line, the function f can be approximated as

$$f \approx \rho_1 E_1 + \rho_3 E_3,$$

which implies that there exists a $[1, 3]$-soliton for $y \gg 0$. The ratio ρ_3/ρ_1 can be used to choose a specific location of this soliton.

Next consider the case of $y \ll 0$. The dominant exponential corresponds to the *least* value of η_j for any given value of c. For large positive c ($\Psi \approx \frac{\pi}{2}$, i.e. $x \to \infty$), E_3 is the dominant term. Decreasing the value of c (rotating the line $x = -cy$ clockwise), the dominant term changes to E_2 for $\kappa_2 + \kappa_3 > c > \kappa_1 + \kappa_2$, and E_1 becomes dominant for $c < \kappa_1 + \kappa_2$. Hence, we have for $y \ll 0$

$$w \longrightarrow \begin{cases} \kappa_1 & \text{as} \quad x \to -\infty, \\ \kappa_2 & \text{for} \quad -(\kappa_1 + \kappa_2)y \ll x \ll -(\kappa_2 + \kappa_3)y, \\ \kappa_3 & \text{as} \quad x \to \infty. \end{cases}$$

These transitions of the dominant exponentials are given by

$$E_1 \quad \longrightarrow \quad E_2 \quad \longrightarrow \quad E_3,$$

as x increases. Near the transition regions, the function f can be approximated by

$$f \approx \rho_1 E_1 + \rho_2 E_2 \quad \text{and} \quad f \approx \rho_2 E_2 + \rho_3 E_3,$$

which correspond to a $[1, 2]$-soliton and a $[2, 3]$-soliton, respectively. That is, the region for $y \ll 0$ is divided by those line-solitons. The shape of solution generated by $f = \rho_1 E_1 + \rho_2 E_2 + \rho_3 E_3$ with all $\rho_i = 1$ (i.e. at $t = 0$ three line-solitons meet at the origin) is illustrated via the contour plot in Fig. 1.3. In this figure, one can see that the line-soliton in $y \gg 0$ labeled by $[1, 3]$ is localized along the phase transition line $x + cy = $ constant with direction parameter $c = \kappa_1 + \kappa_3$; two other line-solitons in $y \ll 0$ labeled by $[1, 2]$ and $[2, 3]$ are localized respectively along the phase transition lines with $c = \kappa_1 + \kappa_2$ and $c = \kappa_2 + \kappa_3$. This solution represents a *resonant interaction* of three line-solitons. In terms of the function w, which is a solution of the Burgers equation in the y-direction, this corresponds to a confluence of two shocks (see Sect. 4.7 in [131]). The resonant condition among those three line-solitons is given by

$$\mathbf{K}_{[1,3]} = \mathbf{K}_{[1,2]} + \mathbf{K}_{[2,3]}, \qquad \Omega_{[1,3]} = \Omega_{[1,2]} + \Omega_{[2,3]},$$

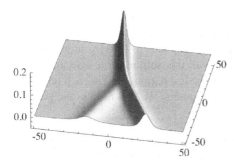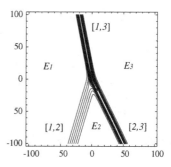

Fig. 1.3 A Y-soliton. Each E_j with $j = 1, 2$ or 3 indicates the dominant exponential term in that region. The boundary of any two adjacent regions gives the line-solitons indicating the transition of the dominant terms E_j. The k-parameters are the same as those in Fig. 1.2, and the line-solitons are determined from the intersection points of the $\eta_j(c)$'s in Fig. 1.2. Here all $\rho_i = 1$ (i.e. $f = E_1 + E_2 + E_3$) so that the three solitons meet at the origin at $t = 0$

which are identically satisfied with $\mathbf{K}_{[i,j]} = (\kappa_j - \kappa_i, \kappa_j^2 - \kappa_i^2)$ and $\Omega_{[i,j]} = -(\kappa_j^3 - \kappa_i^3)$.

The results described in the previous examples can be easily extended to the general case where f has an arbitrary number of exponential terms (see also [17, 86]).

Proposition 1.2 *If* $f = \rho_1 E_1 + \rho_2 E_2 + \cdots + \rho_M E_M$ *with* $\rho_j > 0$ *for* $j = 1, 2, \ldots, M$, *then the solution* u *has the following asymptotic structure:*

(a) *For* $y \gg 0$, *there is only one line-soliton of* $[1, M]$-*type.*
(b) *For* $y \ll 0$, *there are* $M - 1$ *line-solitons of* $[k, k + 1]$-*type for* $k = 1, 2, \ldots,$
    $M - 1$, *counterclockwise from the negative to positive x-axis.*

Figure 1.4 illustrates the evolution of the soliton solution for the case with $M = 4$ and $f = E_1 + E_2 + E_3 + E_4$. In the left figure (at $t = -8$), the middle finite line

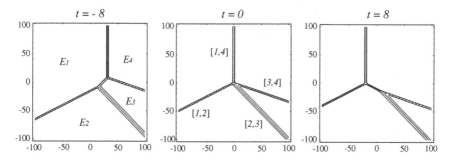

Fig. 1.4 The time evolution of a line-soliton solution for the case with $N = 1$ and $M = 4$. The κ-parameters are $(\kappa_1, \kappa_2, \kappa_3, \kappa_4) = (-2, 0, 1, 2)$. Each E_i in a polyhedral region indicates the dominant exponential in that region, and $[i, j]$ indicates the line-soliton of $[i, j]$-type

corresponds to the line-soliton of [1, 3]-type, and it resonantly interacts with [3, 4]-
and [1, 4]-solitons for $y \gg 0$ and with [1, 2]- and [2, 3]-solitons for $y \ll 0$. In the
right figure (at $t = 8$), the middle finite line corresponds to [2, 4]-soliton, and we
have again two sets of resonant interactions with this soliton. Note also that these
line-solitons divide the xy-plane into four polyhedral regions, and in each region, a
unique exponential function E_i dominates over others. We will discuss the general
case in Chap. 6.

Remark 1.2 It is quite interesting to note that there is a *duality* between a triangulation of polygon and a pattern generated by KP soliton. For example, in the case
of Fig. 1.4, we have a 4-gon with the vertices marked by the index set $\{1, 2, 3, 4\}$,
whose edges may be labeled by [1, 2], [2, 3], [3, 4] and [1, 4]. That is, the edges of
the 4-gon correspond to the asymptotic line-solitons in the KP soliton solution, and
the vertices of 4-gon correspond to the polyhedral regions labeled by the dominant
exponentials. Then the middle (bounded) soliton at $t = -8$ gives a diagonal line
connecting the vertices $\{1, 3\}$, which gives a triangulation of the 4-gon, and the diagonal line corresponds to [1, 3]-soliton. Further discussions on the connection can be
found in [38, 57, 73]. See also Problems 3.4 and 8.4.

1.2 The Multi-component Burgers Equation and the τ-Function

We now consider an *extension* of the first order equation $f_x = w f$, the Cole-Hopf
transformation, to a higher order equation with multi-components $\{w_1, \ldots, w_N\}$
given by

$$\partial_x^N f = w_1 \partial_x^{N-1} f + w_2 \partial_x^{N-2} f + \cdots + w_N f \tag{1.9}$$

Let $\{f_i : i = 1, \ldots, N\}$ be a fundamental set of solutions of the N-th order differential equation (1.9). The transformation between the sets $\{f_i : i = 1, \ldots, N\}$
and $\{w_i : i = 1, \ldots, N\}$ may be considered as the multi-component Cole-Hopf
transformation and can be expressed by

$$\begin{pmatrix} f_1 & f_1^{(1)} & \cdots & f_1^{(N-1)} \\ f_2 & f_2^{(1)} & \cdots & f_2^{(N-1)} \\ \vdots & \vdots & \ddots & \vdots \\ f_N & f_N^{(1)} & \cdots & f_N^{(N-1)} \end{pmatrix} \begin{pmatrix} w_N \\ w_{N-1} \\ \vdots \\ w_1 \end{pmatrix} = \begin{pmatrix} f_1^{(N)} \\ f_2^{(N)} \\ \vdots \\ f_N^{(N)} \end{pmatrix} \tag{1.10}$$

where $f_j^{(n)} := \partial_x^n f_j$. Each function f_j is assumed to satisfy the linear equations,

$$\partial_y f = \partial_x^2 f, \qquad \partial_t f = \partial_x^3 f \qquad \text{for} \qquad j = 1, \ldots, N.$$

Then the set $\{w_i : i = 1, \dots, N\}$ satisfies a *multi-component* extension of the Burgers equation in the y-derivative, while the higher flows with respect to the "times" t_n for $n = 3, 4, \dots$ are the symmetries of this coupled system, thus forming an N-component Burgers hierarchy (this can be formulated in terms of the Sato theory; see [53, 100] and Chap. 2). For example, we have the two-component Burgers equation for (w_1, w_2),

$$\partial_y w_1 = 2w_1 \partial_x w_1 + \partial_x^2 w_1 + 2\partial_x w_2 \,,$$
$$\partial_y w_2 = 2w_2 \partial_x w_1 + \partial_x^2 w_2 \,.$$

Note that if $w_2 = 0$, this system is reduced to the Burgers equation.

Applying the Cramer's rule to (1.10), one can find w_j's. In particular, the function w_1 can be written in the form,

$$w_1 = \partial_x \ln \mathrm{Wr}(f_1, \dots, f_N),$$

where $\mathrm{Wr}(f_1, \dots, f_N)$ is the Wronskian of the functions $\{f_i : i = 1, \dots, N\}$ with respect to the x-variable. We call this Wronskian a τ-*function* of the KP equation (we will give a precise definition of the τ-function later),

$$\tau := \mathrm{Wr}(f_1, \dots, f_N) = \begin{vmatrix} f_1 & f_1^{(1)} & \cdots & f_1^{(N-1)} \\ f_2 & f_2^{(1)} & \cdots & f_2^{(N-1)} \\ \vdots & \vdots & \ddots & \vdots \\ f_N & f_N^{(1)} & \cdots & f_N^{(N-1)} \end{vmatrix}. \tag{1.11}$$

We now have the following theorem.

Theorem 1.2 *The τ-function in (1.11) gives a solution of the KP equation. That is, the function*

$$u(x, y, t) = 2\partial_x^2 \ln \tau(x, y, t)$$

satisfies the KP equation.

Proof First note that inserting the formula $u = 2(\ln \tau)_{xx}$ into the KP equation, one can see that if the τ-function satisfies the following bilinear form, then u satisfies the KP equation.

$$-4(\tau \tau_{xt} - \tau_x \tau_t) + (\tau \tau_{xxxx} - 4\tau_x \tau_{xxx} + 3\tau_{xx}^2) + 3(\tau \tau_{yy} - \tau_y^2) = 0. \tag{1.12}$$

We then show that this equation is nothing but an algebraic relation among the determinants, called a *Plücker relation* (see Sect. 1.2.1). To show this, let us express the τ-function in a symbolic form,

$$\tau = \mathrm{Wr}(f_1, \dots, f_N) =: [0, 1, \dots, N-1],$$

where the numbers represent the derivatives of the f-functions. In general, we define $[l_0, l_1, \ldots, l_{N-1}]$ as

$$[l_0, l_1, \ldots, l_{N-1}] := \begin{vmatrix} f_1^{(l_0)} & f_1^{(l_1)} & \cdots & f_1^{(l_{N-1})} \\ f_2^{(l_0)} & f_2^{(l_1)} & \cdots & f_2^{(l_{N-1})} \\ \vdots & \vdots & \ddots & \vdots \\ f_N^{(l_0)} & f_N^{(l_1)} & \cdots & f_N^{(l_{N-1})} \end{vmatrix}, \tag{1.13}$$

where we do *not* assume the ordering $0 \leq l_0 < l_1 < \cdots < l_{N-1}$. Note that the left expression can be considered as an *algebraic* symbol representing the determinant of N vectors in $\mathbb{R}^\infty$. With this notation, the derivatives of the τ-function are expressed as

$$\begin{aligned} \tau_x &= [0, 1, \ldots, N-2, N], \\ \tau_y &= [0, 1, \ldots, N-3, N, N-1] + [0, 1, \ldots, N-2, N+1] \\ &= -[0, 1, \ldots, N-3, N-1, N] + [0, 1, \ldots, N-2, N+1] \\ \tau_{xx} &= [0, 1, \ldots, N-3, N-1, N] + [0, 1, \ldots, N-2, N+1], \quad \ldots \end{aligned}$$

Note here that since $\partial_{t_n} f = \partial_x^n f$, the derivative ∂_{t_n} shifts each index l_i to $l_i + n$. Then the Eq. (1.12) can be written in the following form:

$$\begin{aligned} [0, 1, \ldots, N-2, N-1][0, 1, \ldots, N-3, N, N+1] \\ - [0, 1, \ldots, N-2, N][0, 1, \ldots, N-3, N-1, N+1] \\ + [0, 1, \ldots, N-2, N+1][0, 1, \ldots, N-3, N-1, N] = 0. \end{aligned}$$

In Proposition 1.3, we will show that this is an example of the *Plücker relations*, which generally hold among the determinants $[l_0, l_1, \ldots, l_{N-1}]$. Then, the proof will be completed by Proposition 1.3. We will provide the details of the Plücker relations in Sect. 1.2.1.   □

Remark 1.3 The main construction of the τ-function in a Wronskian form presented here is based on the Sato theory. However, the formula has been found by several authors in somewhat different but equivalent ways. In particular, we note the following papers [28, 43, 84, 85, 116].

The KP solitons can be also expressed in the so-called Grammian form (see e.g. [56, 98]). In [30], the equivalence between the expressions of the Wronskian and the Grammian has been shown (see Problem 1.5).

1.2.1 The Plücker Relations and the τ-Function

Among the symbols $[l_0, l_1, \ldots, l_{N-1}]$ defined by (1.13), we have the following bilinear relations, called the *Plücker relations* (see e.g. [44]).

Proposition 1.3 *Let $\{\alpha_0, \alpha_1, \ldots, \alpha_{N-2}\}$ and $\{\beta_0, \beta_1, \ldots, \beta_N\}$ be a pair of two sets of nonnegative integers, $\alpha_i, \beta_j \in \mathbb{Z}_{\geq 0}$. Then the symbols $[\alpha_0, \ldots, \alpha_{N-2}, \beta_n]$ and $[\beta_0, \ldots, \widehat{\beta_n}, \ldots, \beta_N]$ for $n = 0, 1, \ldots, N$ satisfy the following relation, the Plücker relation,*

$$\sum_{n=0}^{N} (-1)^n [\alpha_0, \ldots, \alpha_{N-2}, \beta_n] [\beta_0, \ldots, \widehat{\beta_n}, \ldots, \beta_N] = 0.$$

Here $[\beta_0, \ldots, \widehat{\beta_n}, \ldots, \beta_N]$ is the ordered set of $\{\beta_0, \ldots, \beta_N\}$ without β_n.

Proof The formula can be derived by applying the Laplace expansion for the following determinant of $2N \times 2N$ matrix, which is obviously zero,

$$\begin{vmatrix} f_1^{(\alpha_0)} & \cdots & f_1^{(\alpha_{N-2})} & f_1^{(\beta_0)} & f_1^{(\beta_1)} & \cdots & f_1^{(\beta_N)} \\ \vdots & \vdots & \vdots & \vdots & \vdots & \ddots & \vdots \\ f_N^{(\alpha_0)} & \cdots & f_N^{(\alpha_{N-2})} & f_N^{(\beta_0)} & f_N^{(\beta_1)} & \cdots & f_N^{(\beta_N)} \\ 0 & \cdots & 0 & f_1^{(\beta_0)} & f_1^{(\beta_1)} & \cdots & f_1^{(\beta_N)} \\ \vdots & \ddots & \vdots & \vdots & \vdots & \ddots & \vdots \\ 0 & \cdots & 0 & f_N^{(\beta_0)} & f_N^{(\beta_1)} & \cdots & f_N^{(\beta_N)} \end{vmatrix} = 0.$$

The Hirota bilinear equation (1.12) can be obtained by taking $\alpha = \{0, 1, \ldots, N-2\}$ and $\beta = \{0, 1, \ldots, N-3, N-1, N, N+1\}$. Thus, Theorem 1.2 shows that the Wronskian determinant (1.11) gives a particular solution of the KP equation. In this book, we consider finite dimensional solutions given by the following choice of the functions $(f_1, \ldots, f_N)$ in (1.10),

$$f_i(x, y, t) = \sum_{j=1}^{M} a_{ij} e^{\theta_j(x, y, t)}, \quad \text{with} \quad \theta_j = \kappa_j x + \kappa_j^2 y + \kappa_j^3 t, \quad (1.14)$$

where $A := (a_{ij})$ is an $N \times M$ matrix. Thus each KP soliton expressed in the form $u = 2\partial_x^2 \ln \mathrm{Wr}(f_1, \ldots, f_N)$ with (1.14) is parametrized by M parameters $(\kappa_1, \ldots, \kappa_M)$ and an $N \times M$ matrix A. In Chap. 4, the matrix A will be identified as a point of the real Grassmannain $\mathrm{Gr}(N, M)$, which is defined as the set of all N-dimensional subspaces spanned by the row vectors of A.

Example 1.1 Consider the case with $N = 2$ and $M = 3$. We take the linearly independent functions f_1 and f_2 in the form,

$$f_i = \sum_{j=1}^{3} a_{ij} E_j, \qquad i = 1, 2.$$

In this case the τ-function in (1.11) can be explicitly given by

$$\tau = \begin{vmatrix} f_1 & f_1^{(1)} \\ f_2 & f_2^{(1)} \end{vmatrix} = \begin{vmatrix} \begin{pmatrix} a_{11} & a_{12} & a_{13} \\ a_{21} & a_{22} & a_{23} \end{pmatrix} \begin{pmatrix} E_1 & \kappa_1 E_1 \\ E_2 & \kappa_2 E_2 \\ E_3 & \kappa_3 E_3 \end{pmatrix} \end{vmatrix}$$

$$= \begin{vmatrix} a_{11} & a_{12} \\ a_{21} & a_{22} \end{vmatrix} E_{1,2} + \begin{vmatrix} a_{11} & a_{13} \\ a_{21} & a_{23} \end{vmatrix} E_{1,3} + \begin{vmatrix} a_{12} & a_{13} \\ a_{22} & a_{23} \end{vmatrix} E_{2,3},$$

where $E_{i,j} = \mathrm{Wr}(E_i, E_j) = (\kappa_j - \kappa_i) E_i E_j$. Let us consider an example with

$$A = \begin{pmatrix} a_{11} & a_{12} & a_{13} \\ a_{21} & a_{22} & a_{23} \end{pmatrix} = \begin{pmatrix} 1 & 0 & -b \\ 0 & 1 & a \end{pmatrix}, \qquad (1.15)$$

where we assume $a, b > 0$. The τ-function is then given by

$$\tau = E_{1,2} + a E_{1,3} + b E_{2,3}.$$

In order to carry out the asymptotic analysis, one needs to consider the dominance in the set $\{\eta_{i,j} = \eta_i + \eta_j : 1 \le i < j \le 3\}$. This can still be done using Fig. 1.2.

For $y \gg 0$, the transitions of the dominant exponentials are given by following transition scheme,

$$E_{1,2} \quad \longrightarrow \quad E_{1,3} \quad \longrightarrow \quad E_{2,3},$$

as c varies from large positive (i.e. $x \to -\infty$) to large negative values (i.e. $x \to \infty$). The boundary between the regions with the dominant exponentials $E_{1,2}$ and $E_{1,3}$ defines the $[2, 3]$-soliton solution, since here the τ-function can be approximated as

$$\tau \approx E_{1,2} + a E_{1,3} = (\kappa_2 - \kappa_1) E_1 \left(E_2 + a \frac{\kappa_3 - \kappa_1}{\kappa_2 - \kappa_1} E_3 \right)$$

$$= 2(\kappa_2 - \kappa_1) E_1 e^{\frac{1}{2}(\theta_2 + \theta_3 - \theta_{23})} \cosh \tfrac{1}{2}(\theta_2 - \theta_3 + \theta_{23}^0),$$

so that we have

$$u = 2\partial_x^2 \ln \tau \approx \tfrac{1}{2}(\kappa_2 - \kappa_3)^2 \operatorname{sech}^2 \tfrac{1}{2}(\theta_2 - \theta_3 + \theta_{23}^0),$$

where the constant θ_{23}^0 is given by

$$\theta_{23}^0 = \ln \frac{\kappa_2 - \kappa_1}{\kappa_3 - \kappa_1} - \ln a \quad \text{i.e.} \quad a = \frac{\kappa_2 - \kappa_1}{\kappa_3 - \kappa_1} e^{-\theta_{23}^0}.$$

The constant a can be used to determine the location of this soliton. A similar computation as above near the transition boundary of the dominant exponentials $E_{1,3}$ and $E_{2,3}$ yields

$$\tau \approx 2(\kappa_3 - \kappa_1) a E_3 e^{\frac{1}{2}(\theta_1 + \theta_2 - \theta_{12}^0)} \cosh \tfrac{1}{2}(\theta_1 - \theta_2 + \theta_{12}^0),$$
$$u \approx \tfrac{1}{2}(\kappa_1 - \kappa_2)^2 \operatorname{sech}^2 \tfrac{1}{2}(\theta_1 - \theta_2 + \theta_{12}^0),$$

where θ_{12}^0 is given by

$$\theta_{12}^0 = \ln \frac{\kappa_3 - \kappa_1}{\kappa_3 - \kappa_2} - \ln \frac{b}{a} \quad \text{i.e.} \quad b = \frac{\kappa_2 - \kappa_1}{\kappa_3 - \kappa_2} e^{-\theta_{12}^0}.$$

This gives [1, 2]-soliton and its location is determined by the constant b and κ_j's.

For $y \ll 0$, there is only one transition, namely

$$E_{2,3} \quad \longrightarrow \quad E_{1,2},$$

as c varies from large positive values (i.e. $x \to \infty$) to large negative values (i.e. $x \to -\infty$). In this case, a [1, 3]-soliton is formed for $y \ll 0$ at the boundary of the dominant exponentials $E_{2,3}$ and $E_{1,2}$. In a similar computation as the previous cases, the soliton solution of this [1, 3]-type is given by

$$u \approx \tfrac{1}{2}(\kappa_1 - \kappa_3)^2 \operatorname{sech}^2 \tfrac{1}{2}(\theta_1 - \theta_3 + \theta_{13}^0),$$

where the phase constant θ_{13}^0 is given by

$$\theta_{13}^0 = \ln \frac{\kappa_2 = \kappa_1}{\kappa_3 - \kappa_2} - \ln b.$$

Note in particular that those phase constants $\theta_{12}^0, \theta_{23}^0$ and θ_{13}^0 satisfy the resonant relation,

$$\theta_{13}^0 = \theta_{12}^0 + \theta_{23}^0,$$

which implies that those three solitons intersect a point on the xy-plane. Figure 1.5 shows this type of resonant solution with the same κ-parameters as those in Fig. 1.3. Note that this figure can be obtained from Fig. 1.3 by changing $(x, y) \to (-x, -y)$ (this represents a *duality* between those solutions, see Sect. 7.1).

Remark 1.4 Figures 1.3 and 1.5 show two types of the resonant solutions with the same set of three line-solitons of [1, 2]-, [2, 3]- and [1, 3]-types. These solutions are generated from the different sizes of the matrix A with $N = 1$ and $N = 2$ with

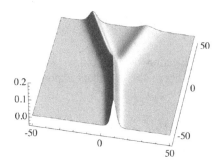 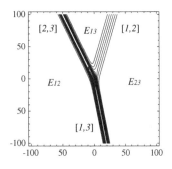

Fig. 1.5 Y-soliton for the case with $N = 2$ and $M = 3$. $E_{i,j}$ indicates the dominant exponential term in this region. The κ-parameters are the same as those in Fig. 1.3. The parameters in the matrix A of (1.15) are chosen as $a = \frac{1}{2}$ and $b = 1$, so that three line-solitons meet at the origin at $t = 0$

the same $M = 3$. In particular, note that the index set of the dominant exponentials around the intersection point is given by $\{1, 2, 3\}$ for the solution in Fig. 1.3 and $\{12, 13, 23\}$ for that in Fig. 1.5.

1.2.2 The Schur Polynomials and the τ-Function

In this section, we introduce the Schur polynomials of the time-variables $\mathbf{t} = (t_1, t_2, \ldots)$ and the corresponding Young diagrams. See [82] for the general information on the Schur polynomials. Then we represent the symbol $[l_0, l_1, \ldots, l_{N-1}]$ as a specific combination of the derivatives of the τ-function in terms of the Schur polynomial. Let us first define the Schur polynomial for the index set $\{l_0, l_1, \ldots, l_{N-1}\}$.

Definition 1.1 Let $p_l(\mathbf{t})$ be the polynomials generated by the identity,

$$\exp\left(\sum_{n=1}^{\infty} k^n t_n\right) = \sum_{l=0}^{\infty} k^l p_l(\mathbf{t}), \tag{1.16}$$

where $\mathbf{t} = (t_1, t_2, \ldots)$. These polynomials are sometimes called the *elementary Schur polynomials*, and they are expressed by

$$p_l(\mathbf{t}) = \sum_{n_1 + 2n_2 + \cdots + ln_l = l} \frac{t_1^{n_1} \cdots t_l^{n_l}}{n_1! \cdots n_l!}.$$

We also set $p_n(\mathbf{t}) = 0$ when $n < 0$. The first few examples are given by

$$p_1 = t_1, \qquad p_2 = t_2 + \frac{1}{2}t_1^2, \qquad p_3 = t_3 + t_1 t_2 + \frac{1}{6}t_1^3.$$

The *Schur polynomial* associated with the index set $\{l_0, l_1, \ldots, l_{N-1}\}$ is defined by the Wronskian with the set of elementary Schur polynomials $\{p_{l_0}, p_{l_1}, \ldots, p_{l_{N-1}}\}$,

$$S_{l_0, l_1, \ldots, l_{N-1}}(\mathbf{t}) = \mathrm{Wr}(p_{l_0}, p_{l_1}, \ldots, p_{l_{N-1}}).$$

Here the derivative in the Wronskian is taken by the t_1-variable.

Note here that we have the following lemma:

Lemma 1.1 *The derivatives of $p_l(\mathbf{t})$ are given by*

$$\frac{\partial p_l}{\partial t_n} = p_{l-n} \quad and \quad p_m = 0 \ \ if m < 0.$$

Proof Taking the derivative of (1.16) with respect to t_n, we have

$$k^n \exp\left(\sum_{m=1}^{\infty} k^m t_m\right) = \sum_{l=0}^{\infty} k^{n+l} p_l(\mathbf{t}) = \sum_{l=0}^{\infty} k^l \frac{\partial p_l}{\partial t_n}(\mathbf{t}).$$

Rearranging the index in the middle term, we have the formula in the lemma. $\square$

The following Corollary is easy to confirm.

Corollary 1.1 *For the index set $I := \{0, 1, \ldots, N-k-1, N-k+i_1, \ldots, N-1+i_k\}$ with $0 \le i_1 \le i_2 \le \cdots \le i_k$, we have*

$$S_I(\mathbf{t}) = S_{i_1, i_2, \ldots, i_k}(\mathbf{t}).$$

That is, the Schur polynomial $S_{l_0, l_1, \ldots, l_N}$ depends only on the nonzero indices in the set $\{l_0, l_1 - 1, \ldots, l_n - n, \ldots, l_{N-1} - (N-1)\}$. For example, we have $S_{0, 1, \ldots, N-1} = S_0 = p_0 = 1$.

Example 1.2 Consider the case $k = 2$. First we note $S_{0,n} = S_{n-1} = p_{n-1}$ for $n = 1, 2, \ldots$. Some of other Schur polynomials are

$$S_{1,2} = \begin{vmatrix} p_1 & p_2 \\ 1 & p_1 \end{vmatrix} = -t_2 + \frac{1}{2}t_1^2, \qquad S_{2,3} = \begin{vmatrix} p_2 & p_3 \\ p_1 & p_2 \end{vmatrix} = -t_1 t_3 + t_2^2 + \frac{1}{12}t_1^4.$$

Remark 1.5 Let us write the variable t_n in the power sum of the variables $(x_1, \ldots, x_m)$

$$t_n = \frac{1}{n} \sum_{j=1}^{m} x_j^n,$$

where m is an arbitrary positive integer. Then we have

$$
\exp\left(\sum_{n=1}^{\infty} k^n t_n\right) = \exp\left(\sum_{n=1}^{\infty}\frac{1}{n}\sum_{j=1}^{m} x_j^n\right) = \exp\left(\sum_{j=1}^{m}\sum_{n=1}^{\infty}\frac{1}{n}(kx_j)^n\right)
$$

$$
= \exp\left(-\sum_{j=1}^{m}\ln(1 - kx_j)\right) = \prod_{j=1}^{m}\frac{1}{1 - kx_j} = \sum_{l=0}^{\infty} h_l(\mathbf{x})k^l.
$$

The functions $h_n(\mathbf{x})$ are the *complete homogeneous symmetric polynomials*, that is, we have

$$
p_l(\mathbf{t}) = h_l(\mathbf{x}) = \sum_{1 \le i_1 \le i_2 \le \cdots \le i_l \le m} x_{i_1} x_{i_2} \cdots x_{i_m}.
$$

Definition 1.2 A Young diagram $Y = \{\lambda_1 \ge \lambda_2 \ge \cdots \ge \lambda_N\}$ is a finite collection of boxes arranged in left-justified rows with the number of boxes λ_j in the j-th row weakly increasing. Then the ordered set $\{l_0, l_1, \ldots, l_{N-1}\}$ with $0 \le l_0 < l_1 < \cdots < l_{N-1}$ can be uniquely expressed by the Young diagram with $\lambda_j = l_{N-j} - (N - j)$ for $j = 1, \ldots, N$. For example,

(a) $\{l_0, l_1, \ldots, l_{N-1}\} = \{0, 1, \ldots, N - 3, N - 2, N - 1\} \equiv \emptyset$ (no boxes)
(b) $\{l_0, l_1, \ldots, l_{N-1}\} = \{0, 1, \ldots, N - 3, N - 2, N\} \equiv \square$
(c) $\{l_0, l_1, \ldots, l_{N-1}\} = \{0, 1, \ldots, N - 3, N - 2, N + 1\} \equiv \square\square$
(d) $\{l_0, l_1, \ldots, l_{N-1}\} = \{0, 1, \ldots, N - 3, N - 1, N + 1\} \equiv \square\!\square$

Then we write the symbol $[l_0, l_1, \ldots, l_{N-1}]$ as τ_Y with the Young diagram $Y = \{\lambda_1 \ge \lambda_2 \ge \cdots \ge \lambda_N\}$ associated to the set $\{l_0, l_1, \ldots, l_{N-1}\}$, i.e.

$$
\tau_Y := [l_0, l_1, \ldots, l_{N-1}] \quad \text{with} \quad \lambda_j = l_{N-j} - (N - j).
$$

The Young diagram associated to $\{l_0, \ldots, l_{N-1}\}$ can be constructed as follows. The southeast boundary of the Young diagram gives a path labeled by increasing numbers from 0, so that the indices $\{l_0, l_1, \ldots, l_{N-1}\}$ appear on the vertical edges in the boundary path as shown below.

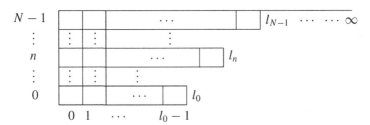

Then we have the following proposition (see e.g. [100, 113, 132]):

Proposition 1.4 *The τ-function τ_Y associated with the Young diagram Y can be expressed by*

$$\tau_Y := [l_0, l_1, \ldots, l_{N-1}] = S_Y(\tilde{\partial})\,\tau,$$

where $\tilde{\partial} := (\partial_1, \frac{1}{2}\partial_2, \frac{1}{3}\partial_3, \ldots)$, and $Y = (\lambda_1, \lambda_2, \ldots, \lambda_N)$ with $\lambda_j = l_{N-j} - (N-j)$.

Proof The result will follow Proposition 1.5 and the orthogonality (1.17) of the Schur polynomials. □

Proposition 1.4 connects the algebraic symbol $[l_0, \ldots, l_{N-1}]$ with a derivative of the τ-function through the Schur polynomial. For example, we have

$$\tau_\emptyset = \tau, \qquad \tau_\square = \tau_x, \qquad \tau_{\square\square} = \frac{1}{2}(\tau_y + \tau_{xx}), \qquad \tau_{\boxminus} = \frac{1}{2}(-\tau_y + \tau_{xx}),$$

$$\tau_{\boxplus} = \frac{1}{3}(-\tau_t + \tau_{xxx}), \qquad \tau_{\boxplus} = \frac{1}{4}\tau_{yy} - \frac{1}{3}\tau_{xt} + \frac{1}{12}\tau_{xxxx}.$$

Then the Plücker relation among the symbols gives the differential equation of the τ-function. In particular, the bilinear Eq. (1.12) can be written in the form,

$$\tau_\emptyset\,\tau_{\boxplus} - \tau_\square\,\tau_{\boxplus} + \tau_{\square\square}\,\tau_{\boxminus} = 0.$$

The main point of the Sato theory is to state that each Plücker relation gives a corresponding member of the KP hierarchy.

For the proof of Proposition 1.4, we first note the following identity.

Proposition 1.5 *The translation of $\tau(\mathbf{t} + \mathbf{a})$ with $\mathbf{a} = (a_1, a_2, \ldots)$ can be expanded in terms of the Schur polynomials,*

$$\tau(\mathbf{t} + \mathbf{a}) = \sum_Y \tau_Y(\mathbf{a}) S_Y(\mathbf{t}),$$

where the sum is over all the sub-Young diagram of the $(\infty)^N$ diagram (i.e. N vertical layers and an infinite number of horizontal layers).

Proof First note that from $\theta(\mathbf{t}; k) = \sum\limits_{n=1}^{\infty} k^n t_n = \sum\limits_{l=0}^{\infty} p_l(\mathbf{t}) k^n$,

$$f_i(\mathbf{a} + \mathbf{t}) = \int_C e^{\theta(\mathbf{a}+\mathbf{t};k)} d\mu(k) = \sum\limits_{l=0}^{\infty} f_i^{(l)}(\mathbf{a}) p_l(\mathbf{t}).$$

Then we have

$$
\tau(\mathbf{a}+\mathbf{t}) = \begin{vmatrix} f_1(\mathbf{a}+\mathbf{t}) & f_2(\mathbf{a}+\mathbf{t}) & \cdots & f_N(\mathbf{a}+\mathbf{t}) \\ f_1'(\mathbf{a}+\mathbf{t}) & f_2'(\mathbf{a}+\mathbf{t}) & \cdots & f_N'(\mathbf{a}+\mathbf{t}) \\ \vdots & \vdots & \ddots & \vdots \\ f_1^{(N-1)}(\mathbf{a}+\mathbf{t}) & f_2^{(N-1)}(\mathbf{a}+\mathbf{t}) & \cdots & f_N^{(N-1)}(\mathbf{a}+\mathbf{t}) \end{vmatrix}
$$

$$
= \left| \begin{pmatrix} f_1 & f_1^{(1)} & f_1^{(2)} & \cdots\cdots \\ f_2 & f_2^{(1)} & f_2^{(2)} & \cdots\cdots \\ \vdots & \vdots & \vdots & \ddots\ddots \\ f_N & f_N^{(1)} & f_N^{(2)} & \cdots\cdots \end{pmatrix} \begin{pmatrix} 1 & 0 & \cdots\cdots & 0 \\ p_1 & 1 & 0 & \cdots & 0 \\ p_2 & p_1 & 1 & \cdots & 0 \\ \vdots & \vdots & \ddots & \ddots & \vdots \end{pmatrix} \right|
$$

Using the Binet-Cauchy Lemma, we obtain the result.   □

Then Proposition 1.4 is a consequence of the orthogonality,

$$
\langle S_Y(\tilde{\mathbf{t}}), S_{Y'}(\mathbf{t}) \rangle := S_Y(\tilde{\partial})\, S_{Y'}(\mathbf{t})\Big|_{t=0} = \delta_{Y,Y'}, \tag{1.17}
$$

where $\tilde{\mathbf{t}} = (t_1, \frac{1}{2}t_2, \frac{1}{3}t_3, \ldots)$ (see Remark 1.6 for the details, also see [100, 113, 132]).

Remark 1.6 Let $\mathbb{C}[[t_1, t_2, \ldots]]$ be the set of formal power series ring on $\mathbb{C}^\infty$. Then we have a gradation,

$$
\mathbb{C}[[t_1, t_2, \ldots]] = \bigoplus_{l \geq 0} V_l^P,
$$

where the graded vector space V_l^P is defined by

$$
V_l^P := \mathrm{Span}_{\mathbb{C}} \left\{ \frac{t_1^{n_1} t_2^{n_2} \cdots t_l^{n_l}}{n_1! \cdots n_l!} : n_1 + 2n_2 + \cdots + ln_l = l \right\}.
$$

Then, each series $f(t) \in \mathbb{C}[[t_1, t_2, \ldots]]$ is given in the form,

$$
f(\mathbf{t}) = \sum_{n_k \in \mathbb{Z}_{\geq 0}} a_{n_1, n_2, \ldots, n_k} \frac{t_1^{n_1} t_2^{n_2} \cdots t_k^{n_k}}{n_1! n_2! \cdots n_k!}.
$$

The coefficients $a_{n_1, n_2, \ldots, n_k}$ can be expressed by

$$
a_{n_1, n_2, \ldots, n_k} = \langle \mathbf{t}^{|n|}, f(\mathbf{t}) \rangle \quad \text{with} \quad \mathbf{t}^{|n|} := t_1^{n_1} t_2^{n_2} \cdots t_k^{n_k}.
$$

Here $\langle \cdot, \cdot \rangle$ is the inner product on $\mathbb{C}[[t_1, t_2, \ldots]]$ defined by

$$
\langle f(\mathbf{t}), g(\mathbf{t}) \rangle := f(\partial) g(\mathbf{t})\Big|_{t=0}.
$$

In particular, we have the orthogonality relation,

$$\left\langle \mathbf{t}^{|n|}, \frac{1}{m}\mathbf{t}^{|m|} \right\rangle = \prod_j \delta_{n_j,m_j} \quad \text{with} \quad \frac{1}{m}\mathbf{t}^{|m|} := \frac{t_1^{m_1} t_2^{m_2} \cdots}{m_1! m_2! \cdots}.$$

The power series ring $\mathbb{C}[[t_1, t_2, \ldots]]$ can also be graded in terms of the Schur polynomials. Let V_l^S be defined by the following vector space of the Schur polynomials,

$$V_l^S := \operatorname{Span}_{\mathbb{C}} \{S_Y(\mathbf{t}) : |Y| = l\}$$

Then there is an isomorphism between those vector spaces V_l^P and V_l^S, which can be given as follows: Expansion of $S_Y(\mathbf{t})$ in the basis of V_l^P is given by

$$S_Y(\mathbf{t}) = \sum_{n_1 + 2n_2 + \cdots k n_k = |Y|} \chi_Y(1^{n_1} 2^{n_2} \cdots k^{n_k}) \frac{t_1^{n_1} t_2^{n_2} \cdots t_k^{n_k}}{n_1! n_2! \cdots n_k!},$$

where the coefficient $\chi_Y(1^{n_1} \cdots k^{n_k})$ is the character of $S_{|Y|}$, which is calculated by

$$\chi_Y(1^{n_1} \cdots l^{n_l}) = \langle \partial^{|n|}, S_Y(\mathbf{t}) \rangle.$$

Then the inversion can be expressed as

$$t_1^{n_1} t_2^{n_2} \cdots t_l^{n_l} = \frac{1}{1^{n_1} 2^{n_2} \cdots l^{n_l}} \sum_{|Y| = n_1 + 2n_2 + \cdots l n_l} \chi_Y(1^{n_1} 2^{n_2} \cdots l^{n_l}) S_Y(\tilde{\mathbf{t}}).$$

where $\tilde{\mathbf{t}} = (t_1, \frac{1}{2}t_2, \frac{1}{3}t_3, \ldots)$. For example, in the cases of $|Y| = 3$ and $|Y| = 4$, the character tables $\chi_Y(1^{n_1} \cdots k^{n_k})$ are given by (see e.g. [45]):

	1^3 $(t_1^3/3!)$	$1 \cdot 2$ $(t_1 t_2)$	3 (t_3)
$S_{\square\square\square}$	1	1	1
$S_{\square\!\square}$	2	0	-1
$S_{\square}$	1	-1	1

	1^4 $(t_1^4/4!)$	$1^2 \cdot 2$ $(t_1^2 t_2/2)$	2^2 $(t_2^2/2)$	$1 \cdot 3$ $(t_1 t_3)$	4 (t_4)
$S_{\square\square\square\square}$	1	1	1	1	1
$S_{\square\square\!\square}$	3	1	-1	0	-1
$S_{\square\square}$	2	0	2	-1	0
$S_{\square\!\square}$	3	-1	-1	0	1
$S_{\square}$	1	-1	1	1	-1

The main theorem of Sato theory of the KP hierarchy may be stated as

Theorem 1.3 *The set of the τ-functions $\{\tau_Y : Y = $ Young diagram $\subset (\infty)^N\}$ provides the Plücker coordinates of the Grassmannian $Gr(N, \infty)$.*

Note that each Plücker relation gives a differential equation with the multi-time variable $\mathbf{t} = (t_1, t_2, \ldots)$. Theorem 1.3 together with Theorem 1.2 gives a definition of the τ-function of the KP hierarchy:

Definition 1.3 (Sato's τ-function) A function $\phi(\mathbf{t})$ of multi-time variable $\mathbf{t} = (t_1, t_2, \ldots)$ is a τ-function of the KP hierarchy if it has an expansion form

$$\phi(\mathbf{t}) = \sum_Y C_Y S_Y(\mathbf{t}),$$

where C_Y are the Plücker coordinates. Note in particular that any Schur polynomial $S_Y(\mathbf{t})$ gives a (*rational*) solution of the KP hierarchy.

Problems

1.1 Derive the following formulas of the Burgers hierarchy by calculating the compatibility of the f-equations,

$$\frac{\partial w}{\partial t_n} = \frac{\partial}{\partial x}\left(\frac{\partial}{\partial x} + w\right)^n 1 \qquad \text{for} \qquad n = 2, 3, \ldots.$$

1.2 Prove Proposition 1.2 for the solution $u = 2\partial_x^2 \ln f$ with $f = \rho_1 E_1 + \cdots + \rho_M E_M$.

1.3 Find the system of equations for (w_1, w_2) given by the compatibility of the following two linear equations,

$$f_{xx} = w_1 f_x + w_2 f, \qquad f_y = f_{xx},$$

which is the two-component Burgers equation. Construct some solutions (w_1, w_2) and discuss the properties.

1.4 Show directly that $\tau = \sum_{1 \le i < j \le 4} \Delta_{ij}(\kappa_j - \kappa_i) E_i E_j$ satisfies the Hirota-bilinear equation (1.12), if and only if Δ_{ij} are the Plücker coordinates, that is, Δ_{ij} satisfy the Plücker relation $\Delta_{12}\Delta_{34} - \Delta_{13}\Delta_{24} + \Delta_{14}\Delta_{23} = 0$.

1.5 It is known that the τ-function can be also expressed in the *Grammian* form (see e.g. [56]),

$$\tau(x, y, t) = \det[I + CF(x, y, t)],$$

where C is an $N \times (M - N)$ constant matrix and F is an $(M - N) \times N$ matrix function whose entries $F_{i,j}(x, y, t)$ are given by

$$F_{i,j}(x, y, t) = \int^x f_i(x, y, t) g_j(x, y, t)\, dx,$$

where $f_i = e^{\phi(p_i)}$ and $g_j = e^{-\phi(q_j)}$ with $\phi(k) = kx + k^2 y + k^3 t$.

Find the κ-parameters and the $N \times M$ matrix A in (1.11) with the functions $\{f_i\}$ in (1.14) in terms of the parameters $\{p_1, p_2, \ldots, p_{M-N}; q_1, q_2, \ldots, q_N\}$ and the coefficient matrix C, so that this τ-function is equivalent to the Wronskian form (1.11) (see [30]).

1.6 Consider the case with $M = 2N$ for the τ-function (1.11) with $\{f_i : i = 1, \ldots, N\}$ given by (1.14). Find a proper choice of the κ-parameters and the matrix $A = (a_{ij})$ in (1.14) so that $u = 2\partial_x^2 \ln \tau$ is the N-soliton solution of the KdV equation, $u_t + 6uu_x + u_{xxx} = 0$. (See [69].)

1.7 Consider the set of N functions $\{f_i(\mathbf{t}) : i = 1, \ldots, N\}$ where $f_i(\mathbf{t})$ is given by

$$f_i(\mathbf{t}) = \sum_{j=1}^{M} a_{i,j} e^{\theta_j(\mathbf{t})} \qquad \text{with} \qquad \theta_j = \sum_{k=1}^{\infty} \kappa_j^k t_k.$$

The τ-function associated with $\{f_i\}$ is given by

$$\tau(\mathbf{t}) = \mathrm{Wr}(f_1, \ldots, f_N).$$

Fix a Young diagram $Y = (\lambda_1, \ldots, \lambda_N)$. Then find the coefficients $\{a_{i,j}\}$ so that we have

$$\tau(\mathbf{t}) \quad \longrightarrow \quad S_Y(\mathbf{t}),$$

in the limit $(\kappa_1, \ldots, \kappa_M) \to (0, \ldots, 0)$. Here $S_Y(\mathbf{t})$ is the Schur polynomial associated to the Young diagram Y. For example (see [23]), if Y is the $N \times (M - N)$ rectangular form, then first choose $a_{i,j} = \kappa_j^{i-1} \rho_j$, which corresponds to the case with $f_i = \partial_x^{i-1} f_1$ and $a_{1,j} = \rho_j$. Then take $\{\rho_j : j = 1, \ldots, M\}$ as the solution of

$$\begin{pmatrix} 1 & 1 & \cdots & 1 \\ \kappa_1 & \kappa_2 & \cdots & \kappa_M \\ \vdots & \vdots & \ddots & \vdots \\ \kappa_1^{M-1} & \kappa_2^{M-1} & \cdots & \kappa_M^{M-1} \end{pmatrix} \begin{pmatrix} \rho_1 \\ \rho_2 \\ \vdots \\ \rho_M \end{pmatrix} = \begin{pmatrix} 0 \\ 0 \\ \vdots \\ 1 \end{pmatrix}.$$

Now the limit $\kappa_j \to 0$ for all j gives the result.

1.8 Let τ be a polynomial of $\mathbf{t} = (t_1, t_2, \ldots)$, which is monic in t_1 of degree N, and write it in the factorization,

$$\tau(\mathbf{t}) = \prod_{i=1}^{N} (t_1 - x_i) \qquad \text{with} \qquad x_i = x_i(t_2, t_3, \ldots). \tag{1.18}$$

Let $p_i = \frac{1}{2} \frac{\partial x_i}{\partial t_2}$. Then, it was shown in [118] (see also [10, 77]) that τ is a τ-function of the KP hierarchy if and only if the N particle system $\{(x_i, p_i) : n = 1, \ldots, N\}$

satisfies the Hamiltonian equations,

$$\frac{\partial x_i}{\partial t_n} = \frac{\partial H_n}{\partial p_i}, \qquad \frac{\partial p_i}{\partial t_n} = -\frac{\partial H_n}{\partial x_i}, \qquad \text{for} \quad n = 2, 3, \ldots, \tag{1.19}$$

where $H_n = (-1)^n \text{tr}(Y^n)$, the trace of the matrix Y^n, and Y is the $N \times N$ matrix defined by

$$Y_{i,i} = p_i, \qquad \text{and} \qquad Y_{i,j} = \frac{1}{x_i - x_j} \quad \text{if} \quad i \neq j.$$

In [21, 94], the following N particle system was shown to be integrable (known as *the Calogero-Moser system*),

$$H = \frac{1}{2} \sum_{i=1}^{N} p_i^2 + \sum_{i<j} \frac{2}{(x_i - x_j)^2}.$$

Find the τ-function (1.18), so that the system (1.19) gives the Calogero-Moser system for the t_2-flow (see [118]).

Chapter 2
Lax-Sato Formulation of the KP Hierarchy

Abstract In this chapter, we briefly review the Lax formulation of the KP hierarchy, which consists of an infinite set of linear equations whose compatibility conditions give rise to the flows corresponding to the KP hierarchy. The main purpose of this section is to highlight the basic framework of integrability underlying the KP theory. Then we show that the multi-component Burgers hierarchy discussed in the previous chapter appears as a finite reduction in the Sato theory. In particular, we emphasize the importance of the τ-function and explain the central role of the τ-function in the KP hierarchy. The materials discussed in this chapter can also be found in [27, 35, 37, 91, 100, 111–113, 132].

2.1 Lax Formulation of the KP Equation

Let L be a pseudo-differential operator of order one defined by

$$L = \partial + u_1 + u_2\partial^{-1} + u_3\partial^{-2} + \cdots,$$

where the coefficients $u_i = u_i(\mathbf{t})$ depend on infinitely many variables $\mathbf{t} = (t_1, t_2, t_3, \ldots)$. The symbol ∂ is a differential operator whereas ∂^{-1} is a formal integration, satisfying $\partial\partial^{-1} = \partial^{-1}\partial = 1$, i.e. ∂^{-1} is a formal inverse of ∂. The operation of ∂^ν with $\nu \in \mathbb{Z}$ follows the generalized Leibnitz rule,

$$\partial^\nu f = \sum_{j \geq 0} \binom{\nu}{j} \partial_x^j(f)\partial^{\nu-j}.$$

For example, we have $\partial f = f_x + f\partial$ and $\partial^{-1}f = f\partial^{-1} - f_x\partial^{-2} + f_{xx}\partial^{-3} - \cdots$ (the latter expression follows from the usual formula of integration by parts). Here we define the *weights* for the functions u_i and ∂^ν as

$$\mathrm{wt}(u_i) = i, \qquad \mathrm{wt}(\partial^\nu) = \nu,$$

© The Author(s) 2017

Y. Kodama, *KP Solitons and the Grassmannians*,

SpringerBriefs in Mathematical Physics 22, DOI 10.1007/978-981-10-4094-8_2

such that the L has a homogeneous weight of one. We also remark that the term u_1 in L can be eliminated by a gauge transformation with a function g such that $u_1 = -g^{-1}\partial_x(g) = -g_x/g$,

$$L \longrightarrow g^{-1}Lg = \partial + \tilde{u}_2\partial^{-1} + \tilde{u}_3\partial^{-2} + \cdots .$$

This can be extended so that we can eliminate *all* u_j with an appropriate gauge operator W, which will be discussed in Sect. 2.2. We will henceforth consider L without the u_1 term,

$$L = \partial + u_2\partial^{-1} + u_3\partial^{-2} + \cdots . \tag{2.1}$$

The Lax form of the KP hierarchy is defined by the infinite set of nonlinear equations

$$\partial_{t_n}(L) = [B_n, L] \quad \text{with} \quad B_n = (L^n)_{\geq 0} \quad n = 1, 2, \ldots, \tag{2.2}$$

where $(L^n)_{\geq 0}$ represents the polynomial part of L^n in ∂, i.e. B_n is a differential operator of order n, and $[B_n, L] := B_nL - LB_n$ is the usual commutator of operators. Since $[B_n, L] = [L^n - (L^n)_{<0}, L] = [L, (L^n)_{<0}]$, each side of the Eq. (2.2) is a pseudo-differential operator of order ≤ -1. Here $(L^n)_{<0}$ is the negative part of L^n in ∂, and note $[\partial, (L^n)_{<0}]$ has no polynomial part. Thus for $n > 1$, each equation in (2.2) is a consistent but infinite system of coupled $(1 + 1)$-evolution equations in t_n and x, for the variables $\{u_i : i = 2, 3, \ldots\}$. The case $n = 1$ yields the equations $\partial_{t_1} u_i = \partial_x u_i$, so we identify t_1 with x. The infinite system (2.2) is compatible, as prescribed by the following theorem:

Theorem 2.1 *The differential operators $B_n = (L^n)_{\geq 0}$ satisfy*

$$\partial_{t_m}(B_n) - \partial_{t_n}(B_m) + [B_n, B_m] = 0. \tag{2.3}$$

Consequently, the flows defined by (2.2) commute i.e., for any $n, m \geq 1$,

$$\partial_{t_n}(\partial_{t_m}(L)) = \partial_{t_m}(\partial_{t_n}(L)).$$

Proof It follows from (2.2) that $\partial_{t_m}(L^n) = [B_m, L^n]$. Then using the decomposition $L^n = B_n + (L^n)_{<0}$, we have

$$\partial_{t_m}(L^n) - \partial_{t_n}(L^m) = [B_m, L^n] - [B_n, L^m]$$
$$= [B_m, B_n] - [(L^m)_{<0}, (L^n)_{<0}].$$

The differential part (≥ 0) of the above equation gives (2.3).

To prove the commutability of the flows, we compute using (2.2) once again

$$\partial_{t_m}(\partial_{t_n}(L)) - \partial_{t_n}(\partial_{t_m}(L)) = [\partial_{t_m}(B_n), L] + [B_n, \partial_{t_m}(L)] - [\partial_{t_n}(B_m), L] - [B_m, \partial_{t_n}(L)]$$
$$= [\partial_{t_m}(B_n) - \partial_{t_n}(B_m), L] + [B_n, [B_m, L]] - [B_m, [B_n, L]].$$

Applying the Jacobi identity for commutators, the right hand side of the above equations becomes $[\partial_{t_m}(B_n) - \partial_{t_n}(B_m) + [B_n, B_m], L]$, which vanishes due to (2.3). □

Equation (2.3) are called the *Zakharov-Shabat* equations for the KP hierarchy. Note that given pair (n, m) with $n > m$, (2.3) gives a system of $n - 1$ equations for $u_2, u_3, \ldots, u_n$, in the variables t_m, t_n and x. For example, consider the case with $n = 3$ and $m = 2$, i.e. $B_2 = (L^2)_{\geq 0} = \partial^2 + 2u_2$ and $B_3 = (L^3)_{\geq 0} = \partial^3 + 3u_2 + 3(u_{2,x} + u_3)$. Then the Zakharov-Shabat equation (2.3) for B_2 and B_3 gives the system

$$\begin{cases} u_{2,t_2} = u_{2,xx} + 2u_{3,x} \\ 2u_{2,t_3} = 3(u_{2,x} + u_3)_{t_2} - (u_{2,xx} - 3u_{3,x} + 3u_2^2)_x \end{cases}$$

After setting $t_2 = y$, $t_3 = t$ and eliminating u_3 from the system, $u = 2u_2$ satisfies the KP equation (1.2).

We also remark that the KP hierarchy (2.2) is given by the compatibility of the linear system

$$\begin{cases} L\phi = k\phi, \\ \partial_{t_n}\phi = B_n\phi, \qquad n = 1, 2, \ldots, \end{cases} \tag{2.4}$$

where $k \in \mathbb{C}$, the spectral parameter, and the eigenfunction $\phi(\mathbf{t}; k)$ with $\mathbf{t} = (t_1, t_2, \ldots)$ will be referred to as the wave function of the KP hierarchy. The compatibility among the second set of equations gives the Zakharov-Shabat equations (2.3).

2.2 The Dressing Transformation

As we mentioned in the previous section, the Lax operator L can be gauge-transformed into the trivial operator ∂, i.e.

$$L \longrightarrow \partial = W^{-1}LW, \tag{2.5}$$

where the operator of the gauge transformation is defined by

$$W = 1 - w_1\partial^{-1} - w_2\partial^{-2} - w_3\partial^{-3} + \cdots. \tag{2.6}$$

The coefficient functions w_i are related to the coefficients u_j of L via the relation $LW = W\partial$ in (2.5), and we have

$$u_2 = w_{1,x}, \quad u_3 = w_{2,x} + w_1w_{1,x}, \quad u_4 = w_{3,x} + (w_1w_2)_x - w_{1,x}^2 + w_1^2w_{1,x}, \quad \cdots$$
$$u_{j+1} = w_{j,x} + F_{j+1}(w_1, w_2, \ldots, w_{j-1}), \quad \cdots,$$

where F_{j+1} are differential polynomials of weight $j + 1$ (note $\mathrm{wt}(w_j) = j$). Thus w_j can be considered as primary variables whose x-derivatives determine the KP variables. The evolutions of w_j with respect to the time variables t_n can be prescribed in a consistent fashion by requiring that the gauge operator W satisfy the following system of equations:

$$\partial_{t_n}(W) = B_n W - W \partial^n \quad \text{for} \quad n = 1, 2, \ldots, \tag{2.7}$$

where B_n is now given by $B_n = (W \partial^n W^{-1})_{\geq 0}$. Notice that this expression for B_n as a differential operator is a consequence of the equations, $[\partial_{t_n}(W) W^{-1}]_{\geq 0} = 0$. Equation (2.7) is sometimes referred to as the Sato equation.

The following theorem asserts that the KP hierarchy is generated by the dressing of the trivial commutation relation $[\partial_{t_n} - \partial^n, \partial] = 0$ by the operator W. Because of this result, the (inverse) gauge transformation, $\partial \to L$, is called the *dressing* transformation for the KP hierarchy, and W is sometimes called the dressing operator.

Theorem 2.2 *If the operator W satisfies the Sato equation (2.7), then the operator $L = W \partial W^{-1}$ satisfies the Lax equation (2.2) for the KP hierarchy, and the operators $B_n = (W \partial^n W^{-1})_{\geq 0} = (L^n)_{\geq 0}$ satisfy the Zakharov-Shabat equations (2.3).*

Proof First, a direct calculation using $L = W \partial W^{-1}$ and $L^n = W \partial^n W^{-1} = B_n + (L^n)_{<0}$ shows that

$$W(\partial_{t_n} - \partial^n) W^{-1} = \partial_{t_n} - \partial_{t_n}(W) W^{-1} - W \partial^n W^{-1}$$
$$= \partial_{t_n} - (\partial_{t_n}(W) + W \partial^n) W^{-1} = \partial_{t_n} - B_n,$$

where the last equality is due to (2.7). Then Eqs. (2.2) and (2.3) follow from the commutator relations

$$0 = W[\partial_{t_n} - \partial^n, \partial] W^{-1} = \partial_{t_n}(L) - [B_n, L],$$
$$0 = W[\partial_{t_n} - \partial^n, \partial_{t_m} - \partial^m] W^{-1} = [\partial_{t_n} - B_n, \partial_{t_m} - B_m],$$

which give the desired formulas. $\square$

It follows from Theorem 2.2 that the flows defined by the Sato equation (2.7) are commutative, i.e. they satisfy the compatibility condition $\partial_{t_m}(\partial_{t_n}(W)) = \partial_{t_n}(\partial_{t_m}(W))$. Indeed, the compatibility condition for (2.7) is equivalent to $[\partial_{t_n} + (L^n)_{<0}, \partial_{t_m} + (L^m)_{<0}] = 0$. Using $L^n = B_n + (L^n)_{<0}$, the commutator term on the left hand side can be decomposed as $[\partial_{t_n} - B_n, \partial_{t_m} - B_m] + [\partial_{t_n} - B_n, L^m] - [\partial_{t_m} - B_m, L^n]$, which vanish due to Theorem 2.2. Finally we note that the KP linear system (2.4) is obtained by the dressing action: $\phi = W \phi_0$ where the (vacuum) wave function ϕ_0 satisfies the *bare* linear system

$$\begin{cases} \partial \phi_0 = k \phi_0, \\ \partial_{t_n} \phi_0 = \partial^n \phi_0 = k^n \phi_0, \quad n = 1, 2, \ldots. \end{cases} \tag{2.8}$$

This equation together with the Sato equation (2.7) forms the basic ingredients of the dressing transformation. We will use the vacuum wave function in the normalized form,

$$\phi_0(\mathbf{t}; k) = e^{\theta(\mathbf{t};k)} \quad \text{with} \quad \theta(\mathbf{t}; k) = \sum_{n=1}^{\infty} k^n t_n. \tag{2.9}$$

2.3 Wave Function ϕ and the τ-Function

The τ-function introduced in Sect. 1.2 plays a fundamental role in the Sato theory of the KP hierarchy. In this section we demonstrate explicitly how the τ-function is related to the dressing operator W, which satisfies the Sato equation (2.7). We restrict our discussion to a finite truncation of the infinite order pseudo-differential operator W for simplicity only, while capturing the flavor of the general theory. Let us then consider the dressing operator for a finite N,

$$W = 1 - w_1 \partial^{-1} - w_2 \partial^{-2} - \cdots - w_N \partial^{-N},$$

and define the differential operator,

$$W_N := W \partial^N = \partial^N - w_1 \partial^{N-1} - w_2 \partial^{N-2} - \cdots - w_N.$$

The equation $W_N f = 0$ gives (1.9) in Sect. 1.2. Since W satisfies the Sato equation, the operator W_N also satisfies

$$\partial_{t_n}(W_N) = B_n W_N - W_N \partial^n \quad \text{with} \quad B_n = (W_N \partial^n W_N^{-1})_{\geq 0}.$$

The following proposition establishes the compatibility conditions leading to the multi-component Burgers hierarchy introduced in Sect. 1.2.

Proposition 2.1 *The N-th order differential equation $W_N f = 0$ is invariant under any flow of the linear heat hierarchy, $\{\partial_{t_n} f = \partial_x^n f : n = 1, 2, \ldots\}$.*

Proof It suffices to show that $\partial_{t_n}(W_N f) = 0$. A direct computation shows

$$\begin{aligned}
\partial_{t_n}(W_N f) &= \partial_{t_n}(W_N) f + W_N \partial_{t_n} f \\
&= (B_n W_N - W_N \partial^n) f + W_N \partial_{t_n} f \\
&= B_n(W_N f) + W_N \left(\partial_{t_n} f - \partial_x^n f\right) = 0.
\end{aligned}$$

Then the desired result follows from the uniqueness of the differential equation. $\square$

Proposition 2.1 provides the compatible system considered in Sect. 1.2,

$$
\begin{cases}
W_N f = 0, \\
\partial_{t_n} f = \partial_x^n f, \quad n = 1, 2, \dots.
\end{cases}
\tag{2.10}
$$

Furthermore, a set $\{f_j : j = 1, 2, \dots, N\}$ of linearly independent solutions of $W_N f = 0$ in (1.9) can be employed to explicitly construct the coefficient functions w_i of the dressing operator W in the form,

$$
w_i = \frac{(-1)^{i+1}}{\tau}
\begin{vmatrix}
f_1 & \cdots & f_1^{(N-i-1)} & f_1^{(N-i+1)} & \cdots & f_1^{(N)} \\
f_2 & \cdots & f_2^{(N-i-1)} & f_2^{(N-i+1)} & \cdots & f_2^{(N)} \\
\vdots & & \vdots & \vdots & & \vdots \\
f_N & \cdots & f_N^{(N-i-1)} & f_N^{(N-i+1)} & \cdots & f_N^{(N)}
\end{vmatrix},
\tag{2.11}
$$

with $\tau = \mathrm{Wr}(f_1, \dots, f_N)$ (see Sect. 1.2). Note that since the t_n-dependence of the w_i is given via the evolution equations $\partial_{t_n} f_j = \partial_x^n f_j$ for $j = 1, 2, \dots, N$, one also has an explicit solution of the Sato equation (2.7).

The formula (2.11) of the coefficients w_i can be exploited to obtain an elegant expression for the wave function ϕ via the τ-function (see also e.g. [9, 37, 91, 132]).

Proposition 2.2 *The wave function ϕ of the linear system (2.4) can be expressed as*

$$
\phi(\mathbf{t}; k) = \frac{\tau(\mathbf{t} - [k^{-1}])}{\tau(\mathbf{t})} \phi_0(\mathbf{t}; k),
\tag{2.12}
$$

where $\phi_0 = e^{\theta(\mathbf{t};k)}$ with $\theta(\mathbf{t}; k) = \sum_{n=1}^{\infty} k^n t_n$, and

$$
(\mathbf{t} - [k^{-1}]) := \left(t_1 - \frac{1}{k}, \ t_2 - \frac{1}{2k^2}, \ t_3 - \frac{1}{3k^3}, \dots \right).
$$

Proof First note that using (2.11), the wave function ϕ can be written as

$$
\phi = W_N \phi_0 = \left(1 - \frac{w_1}{k} - \frac{w_2}{k^2} - \cdots - \frac{w_N}{k^N} \right) \phi_0
$$

$$
= \frac{1}{\tau}
\begin{vmatrix}
f_1 & f_1^{(1)} & \cdots & f_1^{(N)} \\
\vdots & \vdots & \ddots & \vdots \\
f_N & f_N^{(1)} & \cdots & f_N^{(N)} \\
k^{-N} & k^{-N+1} & \cdots & 1
\end{vmatrix}.
$$

Using the elementary column operations, the determinant in the numerator of the above expression can be re-written as

$$\frac{(-1)^N}{k^N} \left| \left(f_i^{(j)} - k f_i^{(j-1)} \right)_{1 \le i, j \le N} \right|.$$

From the integral representation of the functions f_i,

$$f_i(\mathbf{t}) = \int_C e^{\theta(\mathbf{t};\lambda)} \rho_i(\lambda) \, d\lambda \quad \text{for} \quad i = 1, 2, \ldots, N,$$

each matrix element in this determinant is given by

$$f_i^{(j)}(\mathbf{t}) - k f_i^{(j-1)}(\mathbf{t}) = -k \int_C \lambda^{j-1} \left(1 - \frac{\lambda}{k} \right) e^{\theta(\mathbf{t};\lambda)} \rho_i(\lambda) \, d\lambda$$

$$= -k \int_C \lambda^{j-1} e^{-\sum\limits_{n=1}^{\infty} \frac{\lambda^n}{nk^n}} e^{\theta(\mathbf{t};\lambda)} \rho_i(\lambda) \, d\lambda$$

$$= -k e^{-\sum\limits_{n=1}^{\infty} \frac{1}{nk^n} \partial_{t_n}} f_i^{(j-1)}(\mathbf{t}) = -k f_i^{(j-1)} \left(\mathbf{t} - [k^{-1}] \right),$$

where we have used $\ln(1 - \frac{\lambda}{k}) = -\sum\limits_{n=1}^{\infty} \frac{\lambda^n}{nk^n}$. This completes the proof. $\square$

Since the expression of ϕ in Proposition 2.2 does not explicitly depend on N, this formula holds in the general case of the full untruncated version of the operator W (see also [112, 113]). Expanding this formula with respect to k, we have an explicit formula for w_i in (2.11),

$$w_i = -\frac{1}{\tau} p_i(-\tilde{\partial})\tau, \tag{2.13}$$

where $\tilde{\partial} := (\partial_{t_1}, \frac{1}{2}\partial_{t_2}, \frac{1}{3}\partial_{t_3}, \ldots)$ and $p_n(\mathbf{z})$'s are the elementary Schur polynomials in (1.16) (see Problem 2.2). Note that for an N-truncated operator W_N (i.e. $w_n = 0$ if $n > N$) the τ-function satisfies the constraints

$$p_n(-\tilde{\partial})\tau = 0 \quad \text{for} \quad n > N. \tag{2.14}$$

Proposition 2.2 will now be used to derive a set of first integrals of the KP hierarchy that will prove to be useful in our classification of the line-soliton solutions in Chap. 6. The integrability of the KP hierarchy may be demonstrated by the existence of an infinite number of conservation laws in the form,

$$\partial_{t_n} h_j = \partial_x g_{j,n},$$

for some conserved densities h_j and the corresponding conserved fluxes $g_{j,n}$ for any $j, n \geq 1$. These functions are differential polynomials of u_i's in the Lax operator L, and they can be found as follows: Differentiating the quantity $\phi^{-1}\partial_x\phi$ with respect to t_n and using the evolution equation $\partial_{t_n}\phi = B_n\phi$, we first derive the conservation law,

$$\partial_{t_n}\left(\phi^{-1}\partial_x\phi\right) = \partial_x\left(\phi^{-1}B_n\phi\right). \tag{2.15}$$

Next we invert (2.1) using the generalized Leibnitz rule to express the differential operator ∂ in $\partial_x\phi$ in terms of L as

$$\partial = L - v_2 L^{-1} - v_3 L^{-2} - \cdots,$$

where v_j's are the *differential polynomials* of u_i's. Then the conserved density $\phi^{-1}\partial_x\phi$ can be written as an infinite series after using $L^n\phi = k^n\phi$, $n \in \mathbb{Z}$,

$$\phi^{-1}\partial_x\phi = k - \frac{v_2}{k} - \frac{v_3}{k^2} - \cdots.$$

Each function v_j is a conserved density of the KP hierarchy. The first few are given by

$$v_2 = u_2, \qquad v_3 = u_3, \qquad v_4 = u_4 + u_2^2, \qquad v_5 = u_5 - 3u_2 u_3 + u_2 u_{2,x}, \qquad \cdots.$$

Note that these can be expressed simply as sums of the derivatives of $w_1 = \partial_x \ln \tau$. Namely, using (2.12), we have

$$\phi^{-1}\partial_x\phi = \partial_x \ln \phi = \partial_x\left[\theta(\mathbf{t}; k) + \ln \tau(\mathbf{t} - [k^{-1}]) - \ln \tau(\mathbf{t})\right]$$

$$= k + \partial_x\left[\exp\left(-\sum_{n=1}^{\infty}\frac{1}{nk^n}\partial_{t_n}\right) - 1\right]\ln\tau(\mathbf{t}) = k + \sum_{n=1}^{\infty}\frac{1}{k^n}\partial_x p_n(-\tilde{\partial})\ln\tau,$$

which leads to

$$v_{n+1} = -\partial_x p_n(-\tilde{\partial})\ln\tau = -p_n(-\tilde{\partial})w_1. \tag{2.16}$$

For example, we have

$$v_2 = \partial_x w_1, \quad v_3 = \frac{1}{2}\left(\partial_{t_2} - \partial_x^2\right)w_1, \quad v_4 = \frac{1}{3}\left(\partial_{t_3} - \frac{3}{2}\partial_x\partial_{t_2} + \frac{1}{2}\partial_x^3\right)w_1, \cdots$$

$$v_{n+1} = \frac{1}{n}\left(\partial_{t_n} + (\text{h.o.d.})\right)w_1 \qquad \cdots,$$

where (h.o.d) indicates the terms including higher powers of the derivatives. Moreover, if the solutions u_i's of the KP hierarchy decrease rapidly to zero as $|x| \to \infty$, one can define the integrals C_n by

$$C_n := \int_{-\infty}^{\infty} v_{n+1}(x, \ldots)\, dx \qquad \text{for} \quad n = 1, 2, \ldots. \tag{2.17}$$

In particular, if the τ-function gives one line-soliton of $[i, j]$-type, i.e. $\tau(\mathbf{t}) = E_i(\mathbf{t}) + a E_j(\mathbf{t})$ with $E_i(\mathbf{t}) = \exp(\theta(\mathbf{t}; \kappa_i))$ for $\kappa_i < \kappa_j$, then from (2.16) the integrals are

$$C_n = -p_n(-\tilde{\partial}) \ln \tau \Big|_{x=-\infty}^{x=\infty} = \frac{1}{n} \left(\kappa_j^n - \kappa_i^n \right). \tag{2.18}$$

Notice here that since $\kappa_i < \kappa_j$, $\tau \approx E_i = e^{\theta(\mathbf{t}; \kappa_i)}$ for $x \ll 0$ and $\tau \approx a E_j = a e^{\theta(\mathbf{t}; \kappa_j)}$ for $x \gg 0$. In general, if there are N line-solitons of $[i_l, j_l]$-type for $l = 1, \ldots, N$, then we have

$$C_n = \sum_{l=1}^{N} \frac{1}{n} \left(k_{j_l}^n - k_{i_l}^n \right),$$

(see Chap. 6 for the proof). This expression is similar to the N-soliton solutions of the KdV equation (see e.g. [136]). Note in particular that for the KP equation, the C_n's are also independent of y. This fact will be used to prove a part of Theorem 6.1 below.

An alternative set of the conserved densities can be found by observing the following tautological equations (see p. 99 in [37]) in the form of the conservation laws,

$$\partial_{t_m}(\partial_x \partial_{t_n} \ln \tau) = \partial_x(\partial_{t_m} \partial_{t_n} \ln \tau). \tag{2.19}$$

That is, the conserved densities obtained from those equations are given by

$$\tilde{v}_{n+1} := \frac{1}{n} \partial_x \partial_{t_n} \ln \tau,$$

and one can show that the integrals $\tilde{C}_n = \int_{-\infty}^{\infty} \tilde{v}_{n+1}\, dx = \int_{-\infty}^{\infty} v_{n+1} = C_n$ for all n.

In Chap. 6, we will discuss the classification problem of the solutions obtained from the τ-function with finite dimensional solutions of the f-equation and will show that the classification is completely characterized by the asymptotic behavior of the τ-function.

2.4 Bilinear Identity of the τ-Function

There is a unified formulation for the KP hierarchy in terms of the τ-function developed in [35] called *bilinear identity* of the τ-function (later generalized in [128], see also Chap. 3). In this section, we briefly summarize this formula (see also [6, 7]).

Let us first recall the Lax-Sato formulation of the KP hierarchy,

$$L = W \partial W^{-1}, \qquad \partial_{t_n}(W) = B_n W - W \partial^n, \qquad (2.20)$$

where $B_n = (W \partial^n W^{-1})_{\geq 0}$. The wave function $\phi = W \phi_0$ with $\phi_0 = e^\theta$ in (2.9) satisfies

$$\begin{cases} L\phi = k\phi, \\ \partial_{t_n}\phi = B_n\phi. \end{cases}$$

The wave function ϕ can be expressed in terms of the τ-function, i.e. (2.12) in Proposition 2.2.

We now define an adjoint system of the Lax pair, denoted by (L^*, B_n^*),

$$\begin{cases} L^*\phi^* = k\phi^*, \\ \partial_{t_n}\phi^* = -B_n^*\phi^*. \end{cases}$$

Here the (formal) adjoint operator is defined as follows:

- $(\partial^\nu)^* = (-1)^\nu \partial^\nu$ for $\nu \in \mathbb{Z}$,
- For the product of two pseud-differential operators A and B, $(AB)^* = B^*A^*$.

Then the adjoint wave function ϕ^* can be written in the form

$$\phi^*(\mathbf{t}; k) = (W^*)^{-1} e^{-\theta(\mathbf{t};k)}, \qquad (2.21)$$

which is derived by taking the adjoint of (2.20). Then one can show the following main theorem [35, 111] (see also [37, 132]).

Theorem 2.3 *The pair $\{\phi(\mathbf{t}; k), \phi^*(\mathbf{t}'; k)\}$ for arbitrary $\mathbf{t}$ and $\mathbf{t}'$ satisfies the bilinear relation, referred to as the bilinear identity*

$$\oint_{C_\infty} \frac{dk}{2\pi i} \phi(\mathbf{t}; k)\phi^*(\mathbf{t}'; k) = 0,$$

where C_∞ is taken to be a large circle in $\mathbb{C}$.

Then one can express ϕ^* in terms of the τ-function (cf. (2.12)),

$$\phi^*(\mathbf{t}; k) = \frac{\tau(\mathbf{t} + [k^{-1}])}{\tau(\mathbf{t})} e^{-\theta(\mathbf{t};k)},$$

and the bilinear identity in Theorem 2.3 has the following form in terms of the τ-function,

$$\oint_{C_\infty} \frac{dk}{2\pi i} \tau(\mathbf{t} - [k^{-1}]) \, \tau(\mathbf{t}' + [k^{-1}]) \, e^{\theta(\mathbf{t}-\mathbf{t}',k)} = 0. \qquad (2.22)$$

Calculating the residue of (2.22), we can derive the KP hierarchy in terms of the τ-function as follows: First set $\mathbf{t} \to \mathbf{t} - \mathbf{y}$ and $\mathbf{t}' \to \mathbf{t} + \mathbf{y}$. Then we have

$$
\begin{aligned}
0 &= \oint \frac{dk}{2\pi i} \, \tau(\mathbf{t} - \mathbf{y} - [k^{-1}])\tau(\mathbf{t} + \mathbf{y} + [k^{-1}])e^{-2\theta(\mathbf{y};k)} \\
&= \oint \frac{dk}{2\pi i} \left(\exp\left[\sum_{n=1}^{\infty} (y_n + \frac{1}{nk^n})D_n \right] \tau(\mathbf{t}) \cdot \tau(\mathbf{t}) \right) \sum_{l=0}^{\infty} p_l(-2\mathbf{y})k^l \\
&= \oint \frac{dk}{2\pi i} \left(\left[e^{\sum y_n D_n} \sum_{m=0}^{\infty} p_m(\tilde{D})k^{-m} \right] \tau(\mathbf{t}) \cdot \tau(\mathbf{t}) \right) \sum_{l=0}^{\infty} p_l(-2\mathbf{y})k^l \\
&= \oint \frac{dk}{2\pi i} \left(\sum_{m=0}^{\infty}\sum_{l=0}^{\infty} \frac{1}{k^{m-l}} p_l(-2\mathbf{y}) p_m(\tilde{D}) e^{\sum y_n D_n} \right) \tau(\mathbf{t}) \cdot \tau(\mathbf{t}) \\
&= \sum_{l=0}^{\infty} p_l(-2\mathbf{y}) p_{l+1}(\tilde{D}) e^{\sum y_n D_n} \tau(\mathbf{t}) \cdot \tau(\mathbf{t}).
\end{aligned}
$$

Here the operator D_n is the Hirota derivative defined by

$$
D_n^m f \cdot g = (\partial_{t_n} - \partial_{s_n})^m f(t_n)g(s_n)\big|_{t_n = s_n},
$$

so that we have

$$
\exp(\alpha D_n) f(t_n) \cdot g(t_n) = f(t_n + \alpha)g(t_n - \alpha).
$$

We have also defined

$$
\tilde{D} := \left(D_1, \frac{1}{2}D_2, \frac{1}{3}D_3, \dots \right).
$$

Then expanding the bilinear equation in terms of $\mathbf{y}$, we can obtain a (Hirota) bilinear equation for the τ-function as the coefficient of each monomial $y_1^{i_1} \cdots y_k^{i_k}$. In particular, we obtain the following equation at the coefficient of y_k,

$$
\left[-2p_{k+1}(\tilde{D}) + D_1 D_k \right] \tau \cdot \tau = 0 \qquad \text{for} \qquad k = 1, 2, \dots. \qquad (2.23)
$$

Here, we note that the equations for $k = 1, 2$ are trivial, and at $k = 3$, we have

$$
\begin{aligned}
0 &= \left[-2p_4(\tilde{D}) + D_1 D_3 \right] \tau \cdot \tau \\
&= \left[-2\left(\frac{1}{4}D_4 + \frac{1}{3}D_1 D_3 + \frac{1}{4}D_1^2 D_2 + \frac{1}{8}D_2^2 + \frac{1}{24}D_1^4 \right) + D_1 D_3 \right] \tau \cdot \tau \\
&= -\frac{1}{12} \left(-4D_1 D_3 + 3D_2^2 + D_1^4 \right) \tau \cdot \tau.
\end{aligned}
$$

which is the KP equation in the τ-function form (1.12).

Remark 2.1 In [55], Hirota found a bilinear difference equation,

$$\left[Z_1 \exp(D_1) + Z_2 \exp(D_2) + Z_3 \exp(D_3)\right] \tau \cdot \tau = 0, \tag{2.24}$$

where Z_i are constants satisfying $Z_1 + Z_2 + Z_3 = 0$. By taking suitable limits, this equation gives several integrable systems including the KP equation [90]. Related to this bilinear equation, one can also construct other formulation of the τ-function (see e.g. [14, 51], also [122] for a further development).

2.5 Hirota Perturbation Method for N-Soliton Solutions

Here we briefly explain how one gets an N-soliton solution from the bilinear form using a *perturbation method* introduced by Hirota (see [56] for a comprehensive review on the method). The Hirota bilinear equation of the KP equation (1.12) is given by

$$P(D_x, D_y, D_t)\tau \cdot \tau := (-4D_x D_t + D_x^4 + 3D_y^2)\tau \cdot \tau = 0. \tag{2.25}$$

Note that the function $P(D_x, D_y, D_t)$ gives the soliton dispersion relation (1.6) (see also [97].),

$$P(K_{[i,j]}^x, K_{[i,j]}^y, \Omega_{[i,j]}) = -4K_{[i,j]}^x \Omega_{[i,j]} + (K_{[i,j]}^x)^4 + 3(K_{[i,j]}^y)^2 = 0.$$

The perturbation method assumes τ in the following N-term expansion with a parameter ε,

$$\tau = 1 + \varepsilon g_1 + \varepsilon^2 g_2 + \cdots + \varepsilon^N g_N.$$

The first few terms in the coefficients of the powers of ε are given by

$$P(D_x, D_y, D_t)1 \cdot 1 = 0,$$
$$P(D_x, D_y, D_t)1 \cdot g_1 = 0,$$
$$P(D_x, D_y, D_t)g_1 \cdot g_1 + 2P(D_x, D_y, D_t)1 \cdot g_2 = 0,$$
$$P(D_x, D_y, D_t)g_1 \cdot g_2 + P(D_x, D_y, D_t)1 \cdot g_3 = 0.$$

The first equation is trivially satisfied. The second equation implies

$$P(\partial_x, \partial_y, \partial_t)g_1 = -4\partial_x \partial_t g_1 + \partial_x^4 g_1 + 3\partial_y^2 g_1 = 0,$$

which is nothing but the linearized KP equation. The solution g_1 can be given by

$$g_1 = \iint \rho_1(p, q)e^{px+qy+\omega t}\,dpdq,$$

where (p, q, ω) satisfies the dispersion relation, $P(p, q, \omega) = -4p\omega + p^4 + 3q^2 = 0$. For one-soliton solution ($N = 1$), we choose

$$g_1 = e^{px+qy+\omega t+c}.$$

Then one can easily show that

$$P(D_x, D_y, D_t)g_1 \cdot g_1 = 0.$$

This implies that $\tau = 1 + \varepsilon g_1$ (i.e. $g_n = 0$ for $n \geq 2$) is an exact solution, and the solution u of the KP equation is given by

$$u = 2\partial_x^2 \ln \tau = \frac{1}{2}p^2 \mathrm{sech}^2 \frac{1}{2}(px + qy + \omega t + c).$$

Note here that the dispersion relation (1.6) has a parametrization,

$$(p, q, \omega) = (\kappa_i - \kappa_j, \kappa_i^2 - \kappa_j^2, \kappa_i^3 - \kappa_j^3),$$

where κ_i and κ_j are arbitrary constants (cf. (1.4)).
  For 2-soliton solution ($N = 2$), we take

$$g_1 = E_1 + E_2 \quad \text{with} \quad E_i := e^{p_i x + q_i y + \omega_i t + c_i},$$

where (p_i, q_i, ω_i) satisfies the dispersion relation for each $i = 1, 2$. Then the third equation at the order ε^2 requires

$$P(D_x, D_y, D_t)E_1 \cdot E_2 + P(D_x, D_y, D_t)1 \cdot g_2 = 0.$$

The solution g_2 can be found in the form

$$g_2 = A_{12}E_1 E_2 \quad \text{with} \quad A_{12} = -\frac{P(p_1 - p_2, q_1 - q_2, \omega_1 - \omega_2)}{P(p_1 + p_2, q_1 + q_2, \omega_1 + \omega_2)}.$$

One can also show that

$$P(D_x, D_y, D_t)g_1 \cdot g_2 = A_{12}P(D_x, D_y, D_t)(E_1 + E_2) \cdot E_1 E_2 = 0,$$
$$P(D_x, D_y, D_t)g_2 \cdot g_2 = A_{12}^2 P(D_x, D_y, D_t)E_1 E_2 \cdot E_1 E_2 = 0.$$

The solution τ is then given by

$$\tau = 1 + e^{\xi_1} + e^{\xi_2} + A_{12}e^{\xi_1 + \xi_2} \quad \text{with} \quad \xi_i = p_i x + q_i y + \omega_i t + c_i',$$

where $c_i' = c_i + \ln \varepsilon$ are arbitrary constants.

Using the dispersion relation (1.6), one can set

$$p_1 = \kappa_1 - \kappa_2, \qquad q_1 = \kappa_1^2 - \kappa_2^2, \qquad \omega_1 = \kappa_1^3 - \kappa_2^3,$$
$$p_2 = \kappa_3 - \kappa_4, \qquad q_2 = \kappa_3^2 - \kappa_4^2, \qquad \omega_2 = \kappa_3^3 - \kappa_4^3,$$

with arbitrary parameters $(\kappa_1, \ldots, \kappa_4)$. Then the τ-function can be written in the form

$$\tau = 1 + e^{\xi_1} + e^{\xi_2} + A_{12} e^{\xi_1 + \xi_2} = \frac{1}{ab E_{2,4}} \left(E_{1,3} + a E_{1,4} + b E_{2,3} + ab E_{2,4} \right),$$

where $E_{i,j} = (\kappa_j - \kappa_i) e^{\theta_i + \theta_j}$ with $\theta_i = \kappa_i x + \kappa_i^2 y + \kappa_i^3 t$ and $A_{12} = \frac{(\kappa_1 - \kappa_3)(\kappa_2 - \kappa_4)}{(\kappa_1 - \kappa_4)(\kappa_2 - \kappa_3)}$. The constants c_i's in ξ_i's are given by

$$c_1' = -\ln \left(b \frac{\kappa_4 - \kappa_2}{\kappa_4 - \kappa_1} \right), \qquad c_2' = -\ln \left(a \frac{\kappa_4 - \kappa_2}{\kappa_3 - \kappa_2} \right),$$

with arbitrary constants a and b. Since $u = 2\partial_x^2 \ln \tau$, the following function can be used for the same solution u,

$$\tilde{\tau} = E_{1,3} + a E_{1,4} + b E_{2,3} + ab E_{2,4},$$

which is given by the Wronskian form for $N = 2$ and $M = 4$ with the 2×4 matrix

$$A = \begin{pmatrix} 1 & b & 0 & 0 \\ 0 & 0 & 1 & a \end{pmatrix}.$$

That is, the $\tilde{\tau}$-function is given by the Wronskian form,

$$\tilde{\tau} = \begin{vmatrix} f_1 & \partial_x f_1 \\ f_2 & \partial_x f_2 \end{vmatrix} \quad \text{with} \quad (f_1, f_2) = (E_1, E_2, E_3, E_4) A^T,$$

where $E_i = e^{\theta_i}$ and A^T is the transpose of the matrix A. Notice that this matrix A is *not* a generic form of the general 2×4 matrix. Thus the solutions obtained by Hirota's perturbation method gives only a *special* class of the KP solitons. Note also that the τ-function generated by the generic 2×4 matrix A contains *six* exponential terms. We will discuss this issue in Chap. 6 where we construct more general KP solitons and classify these solutions based on the Wronskian structure of the τ-function.

Problems

2.1 Let P_2 be the differential operator of order two given by

$$P_2 = \partial^2 + u,$$

where u is a function of multi-variables $\mathbf{t} = (t_1, t_2, \ldots)$.

(a) Find the pseudo-differential operator $L = \partial + u_2\partial^{-1} + u_3\partial^{-2} + \cdots$ in (2.1), so that $L^2 = P_2$. That is, find each u_k in terms of a differential polynomial of u.

(b) Define B_n as in (2.2), i.e. $B_n = (L^n)_{\geq 0}$. Then show that $u_{t_n} = 0$ if n is even, and the Lax equations $\partial_{t_{2k+1}}(L) = [B_{2k+1}, L]$ for $k = 1, 2, \ldots$ give the KdV hierarchy. Note here that the Lax form of the KdV hierarchy can be written as a pair of differential operators (P, B_n), i.e. $\partial(P) = [B_n, P]$, which is the original formulation given in [78].

(c) Discuss the generalization of this procedure. That is, for a given differential operator of order N, $P_N = \partial^N + v_2\partial^{N-2} + v_3\partial^{N-2} + \cdots + v_N$, find L so that $L^N = P_N$, and define $B_n = (L^n)_{\geq 0}$. Then one can construct a hierarchy in the Lax form, $\partial_{t_n}(L) = [B_n, L]$, which is called the *Gel'fand-Dikey* hierarchy. Note that $[B_n, L] = 0$ if $n = kN$ for $k = 1, 2, \ldots$ (see [37]).

2.2 Derive the formula (2.13) of the function w_i in the dressing operator W, and the formula (2.14) for the truncated operator W_N.

2.3 Let ϕ be a particular solution of (2.4), and let W be a dressing operator satisfying the Sato equation (2.7). Then prove the following.

Proposition 2.3 *Let $G = \partial - \phi^{-1}\phi_x$. Then the operator defined by*

$$\widetilde{W} = GW\partial^{-1} = 1 - \tilde{w}_1\partial^{-1} - \tilde{w}_2\partial^{-2} - \cdots$$

also satisfies the Sato equation (2.7), and the new coefficient functions $\tilde{w}_n$ are given by

$$\tilde{w}_1 = w_1 + \phi^{-1}\phi_x, \qquad \tilde{w}_n = w_n + \phi(\phi^{-1}w_{n-1})_x \quad \text{for } n > 1.$$

This defines the *Darboux* transformation for the KP hierarchy, and using this transformation, one can also find the KP solution in the Wronskian form (see [28, 84]).

2.4 Derive the formula ϕ^* in (2.21) and prove Theorem 2.3.

2.5 Identify the Plücker relation for the next member of the KP hierarchy,

$$\left[-2p_4(\tilde{D}) + D_1 D_4\right]\tau \cdot \tau = 0.$$

2.6 First prove the following identity of the bilinear form,

$$p_l(\tilde{D})\tau \cdot \tau = \sum_{i+j=l} \left(p_i(\tilde{\partial})\tau\right)\left(p_j(-\tilde{\partial})\tau\right),$$

where p_l is the elementary Schur polynomial. Then show that the KP hierarchy (2.23) is just a Plücker relation for each k when we express the derivative of τ as $S_Y(\tilde{\partial})\tau = \tau_Y$ with the corresponding Young diagram Y.

Chapter 3
Two-Dimensional Solitons

Abstract In Chap. 2, we have shown that the KP hierarchy admits particular solutions, called the KP solitons, the main subject of this book, which are expressed by the Wronskian form. In this chapter, we show that this determinant structure is common for other two-dimensional integrable systems generated by several reductions of the *modified bilinear identity* proposed by Ueno-Takasaki [128] (see [123] for a further generalization of the bilinear identity). In addition to the KP hierarchy, these integrable systems also include the two-dimensional Toda lattice hierarchy and the Davey-Stewartson hierarchy. Here we construct their soliton solutions in the determinant form and show that their wave parameters for these solutions are chosen from conic curves, that is, the KP soliton from the parabola, the two-dimensional Toda soliton from the hyperbola, and the Davey-Stewartson soliton from the circle.

3.1 The τ-Functions for Two-Dimensional Soliton Equations

In [128], Ueno and Takasaki proposed the *modified* bilinear identity for the two-dimensional Toda lattice hierarchy whose τ-function depends on the variables $(s, \mathbf{x}, \bar{\mathbf{x}})$, denoted by

$$\tau^{(s)}(\mathbf{x}, \bar{\mathbf{x}}) \quad \text{with} \quad s \in \mathbb{C}, \quad \mathbf{x} = (x_n : n \in \mathbb{Z}_{>0}) \text{ and } \bar{\mathbf{x}} = (x_n : n \in \mathbb{Z}_{<0}).$$

The bilinear identity for these τ-functions is expressed by

$$\oint_{C_\infty} \frac{dk}{2\pi i} \tau^{(s')}(\mathbf{x}' - [k^{-1}], \bar{\mathbf{x}}') \, \tau^{(s)}(\mathbf{x} + [k^{-1}], \bar{\mathbf{x}}) k^{s'-s} e^{\theta(\mathbf{x}'-\mathbf{x};k)+\theta(\bar{\mathbf{x}}'-\bar{\mathbf{x}};k^{-1})}$$

$$= \oint_{C_0} \frac{dk}{2\pi i} \tau^{(s'+1)}(\mathbf{x}', \bar{\mathbf{x}}' - [k]) \, \tau^{(s-1)}(\mathbf{x}, \bar{\mathbf{x}} + [k]) k^{s'-s} e^{\theta(\mathbf{x}'-\mathbf{x};k)+\theta(\bar{\mathbf{x}}'-\bar{\mathbf{x}};k^{-1})}$$

© The Author(s) 2017
Y. Kodama, *KP Solitons and the Grassmannians*,
SpringerBriefs in Mathematical Physics 22, DOI 10.1007/978-981-10-4094-8_3

for any $(s, \mathbf{x}, \bar{\mathbf{x}})$ and $(s', \mathbf{x}', \bar{\mathbf{x}}')$. Here the contour C_0 (C_∞) is taken to be a small circle around $k = 0$ $(k = \infty)$. Also recall that

$$\theta(\mathbf{x}; \lambda) = \sum_{n=1}^{\infty} \lambda^n x_n, \qquad \mathbf{x} - [\lambda^{-1}] = \left(x_1 - \frac{1}{\lambda}, x_2 - \frac{1}{2\lambda^2}, \ldots \right), \qquad \text{etc.}$$

This modified bilinear identity is a generalization of the bilinear identity (2.22) for the KP hierarchy.

We now derive some bilinear equations for the τ-functions along the same lines as in the case of the KP hierarchy. We set

$$\begin{aligned} \mathbf{x}' &\to \mathbf{x} - \mathbf{y}, & \mathbf{x} &\to \mathbf{x} + \mathbf{y}, \\ \bar{\mathbf{x}}' &\to \bar{\mathbf{x}} - \bar{\mathbf{y}}, & \bar{\mathbf{x}} &\to \bar{\mathbf{x}} + \bar{\mathbf{y}}. \end{aligned}$$

Then the bilinear equation becomes

$$\oint_{C_\infty} \frac{dk}{2\pi i} \tau^{(s')}(\mathbf{x} - \mathbf{y} - [k^{-1}], \bar{\mathbf{x}} - \bar{\mathbf{y}})\, \tau^{(s)}(\mathbf{x} + \mathbf{y} + [k^{-1}], \bar{\mathbf{x}} + \bar{\mathbf{y}})k^{s'-s}e^{\theta(-2\mathbf{y}, k) + \theta(-2\bar{\mathbf{y}}; k^{-1})}$$

$$= \oint_{C_0} \frac{dk}{2\pi i} \tau^{(s'+1)}(\mathbf{x} - \mathbf{y}, \bar{\mathbf{x}} - \bar{\mathbf{y}} - [k])\, \tau^{(s-1)}(\mathbf{x} + \mathbf{y}, \bar{\mathbf{x}} + \bar{\mathbf{y}} + [k])k^{s'-s}e^{\theta(-2\mathbf{y}, k) + \theta(-2\bar{\mathbf{y}}; k^{-1})},$$

which gives

$$\sum_{n,m \geq 0} p_n(-2\mathbf{y}) p_m(-2\bar{\mathbf{y}}) p_{n-m+s'-s+1}(\tilde{D}_x) e^{\sum y_n D_{x_n}} e^{\sum \bar{y}_n D_{\bar{x}_n}} \tau^{(s)} \cdot \tau^{(s')}$$

$$= \sum_{n,m \geq 0} p_n(-2\mathbf{y}) p_m(-2\bar{\mathbf{y}}) p_{-n+m-s'+s-1}(\tilde{D}_{\bar{x}}) e^{\sum y_n D_{x_n}} e^{\sum \bar{y}_n D_{\bar{x}_n}} \tau^{(s-1)} \cdot \tau^{(s'+1)},$$

where $p_n(\mathbf{x})$'s are the elementary Schur polynomials in (1.16). One can see that the KP hierarchy can be obtained by taking $\bar{\mathbf{y}} = \mathbf{0}$ with $s = s'$. In particular, the coefficient of y_3 gives the bilinear equation of the KP equation as we saw in the previous Chapter. Also, the coefficient of the constant term obtained at $m = n$ and $s' = s + 1$ gives (3.2) below, and at $m = n$ and $s' = s - 3$, gives (3.3). One can also find that the coefficient of $\bar{y}_1$ gives (3.4) below:

$$[D_1(-4D_3 + D_1^3) + 3D_2^2]\tau^{(s)} \cdot \tau^{(s)} = 0, \tag{3.1}$$

$$(D_2 + D_1^2)\tau^{(s)} \cdot \tau^{(s+1)} = 0, \tag{3.2}$$

$$(D_{-2} - D_{-1}^2)\tau^{(s)} \cdot \tau^{(s+1)} = 0, \tag{3.3}$$

$$D_1 D_{-1}\tau^{(s)} \cdot \tau^{(s)} = 2[(\tau^{(s)})^2 - \tau^{(s+1)}\tau^{(s-1)}]. \tag{3.4}$$

Here we write $D_{x_n} = D_n$ and $D_{\bar{x}_n} = D_{-n}$. Equation (3.1) is the KP bilinear equation, and (3.2) is referred to as the Bäcklund transformation of the KP equation between two τ-functions, $\tau = \tau^{(s)}$ and $\tau' = \tau^{(s+1)}$ (see e.g. [56]). The third one (3.3) is an extension of the Bäcklund transformation for the negative flows. Equation (3.4) is the two-dimensional Toda bilinear equation for $s \in \mathbb{Z}$ [128]. It is interesting to note that (3.4) together with (3.2) and (3.3) provides the bilinear form of the Davey-Stewartson equation (see below for the details).

3.2 Soliton Solutions

One can show that for each s, the $\tau^{(s)}$-function in the Wronskian with some functions $\{f_i^{(s)}(\mathbf{x}) : i = 1, \ldots, N\}$ satisfies the Hirota bilinear equation for $\mathbf{x} := (x_k : k \in \mathbb{Z}\backslash\{0\})$,

$$\tau^{(s)}(\mathbf{x}) = \mathrm{Wr}(f_1^{(s)}, \ldots, f_N^{(s)})(\mathbf{x}), \tag{3.5}$$

where the derivatives in the Wronskian are on the x_1-variable, if each $f_i^{(s)}$ satisfies

$$\frac{\partial f_i^{(s)}}{\partial x_k} = f_i^{(s+k)} \quad \text{for} \quad k \in \mathbb{Z}\backslash\{0\}. \tag{3.6}$$

That is, we have the following proposition.

Proposition 3.1 *The set of the τ-functions in (3.5) satisfies the Hirota bilinear equations (3.1)–(3.4).*

The proof is similar to that shown in Theorem 1.2, i.e. the proof is to show that the $\tau^{(s)}$-function satisfies certain Plücker relations (see e.g. [56] for the details and elegant proofs).

Let us take the following exponential functions which are simple solutions of (3.6),

$$E_j^{(s)}(\mathbf{x}) = \lambda_j^s \, e^{\theta_j(\mathbf{x})} \quad \text{with} \quad \theta_j(\mathbf{x}) := \sum_{k \in \mathbb{Z}\backslash\{0\}} \lambda_j^k x_k \tag{3.7}$$

with constant parameters $\lambda_j \in \mathbb{C}$. We then define each $f_i^{(s)}$ as a finite sum of those exponential functions $\{E_j^{(s)} : j = 1, \ldots, M\}$, i.e.

$$f_i^{(s)}(\mathbf{x}) = \sum_{j=1}^M a_{ij} E_j^{(s)}(\mathbf{x}) \quad \text{for} \quad i = 1, 2, \ldots, N, \tag{3.8}$$

with an $N \times M$ matrix $A = (a_{ij})$. It is an easy exercise (using the Binet-Cauchy Lemma) to show that the $\tau^{(s)}$-function can be expressed as

$$\tau^{(s)}(\mathbf{x}) = \sum_{I \in \binom{[M]}{N}} \Delta_I(A) \, E_I^{(s)}(\mathbf{x}), \quad \text{with} \quad E_I^{(s)} = \mathrm{Wr}(E_{i_1}^{(s)}, \ldots, E_{i_N}^{(s)}),$$

for $I = \{i_1, \ldots, i_N\}$, and $\binom{[M]}{N}$ denotes the set of all N-index subsets of $[M] := \{1, 2, \ldots, M\}$. Here $\Delta_I(A)$ is the $N \times N$ minor of the matrix A with the columns labeled by the indices in I, and the Wronskian $E_I^{(s)}$ is

$$
E_I^{(s)} =
\begin{vmatrix}
E_{i_1}^{(s)} & E_{i_1}^{(s+1)} & \cdots & E_{i_1}^{(s+N-1)} \\
E_{i_2}^{(s)} & E_{i_2}^{(s+1)} & \cdots & E_{i_2}^{(s+N-1)} \\
\vdots & \vdots & \ddots & \vdots \\
E_{i_N}^{(s)} & E_{i_N}^{(s+1)} & \cdots & E_{i_N}^{(s+N-1)}
\end{vmatrix}
= \left(\prod_{k=1}^{N} \lambda_{i_k}^s e^{\theta_{i_k}} \right)
\begin{vmatrix}
1 & \lambda_{i_1} & \cdots & \lambda_{i_1}^{N-1} \\
1 & \lambda_{i_2} & \cdots & \lambda_{i_2}^{N-1} \\
\vdots & \vdots & \ddots & \vdots \\
1 & \lambda_{i_N} & \cdots & \lambda_{i_N}^{N-1}
\end{vmatrix}
$$

$$
= \left(\prod_{k=1}^{N} \lambda_{i_k}^s e^{\theta_{i_k}} \right) \prod_{j>k} (\lambda_{i_j} - \lambda_{i_k}) =: \Lambda_I^{(s)} e^{\Theta_I(\mathbf{x})},
$$

where Λ_I and $\theta_I(\mathbf{x})$ for the ordered set $I = \{i_1 < \cdots < i_N\}$ are defined by

$$
\Lambda_I^{(s)} = \left(\prod_{k=1}^{N} \lambda_{i_k}^s \right) \prod_{j>k} (\lambda_{i_j} - \lambda_{i_k}), \quad \text{and} \quad \Theta_I(\mathbf{x}) = \sum_{k=1}^{N} \theta_{i_k}(\mathbf{x}).
$$

Notice that the parameter s appears only in $\Lambda_I^{(s)}$ and the τ-function has a simple form,

$$
\tau^{(s)}(\mathbf{x}) = \sum_{I \in \binom{[M]}{N}} \Delta_I(A) \, \Lambda_I^{(s)} \, e^{\Theta_I(\mathbf{x})}. \tag{3.9}
$$

We construct *real* and *regular* solutions of those two-dimensional soliton equations mentioned in the previous section by choosing appropriate parameters s and λ_j's in the exponential functions in (3.7) and the coefficient matrix A in (3.8).

3.2.1 The KP Solitons

As we discussed in the previous chapters, the bilinear equation (3.1) gives the KP equation through $u(x, y, t) = 2\partial_x^2 \ln \tau^{(s)}(x, y, t)$ with the choice of the coordinates

$$
x = x_1, \qquad y = x_2, \qquad t = x_3.
$$

For real solutions, we choose $s = 0$, $\lambda_j = \kappa_j \in \mathbb{R}$ and a real $N \times M$ matrix A. Then the τ-function is given by

$$
\tau^{(0)}(x, y, t) = \sum_{I \in \binom{[M]}{N}} \Delta_I(A) \, K_I \, e^{\Theta_I(x,y,t)} \quad \text{with} \quad K_I = \prod_{j>k} (\kappa_{i_j} - \kappa_{i_k}).
$$

If we choose the matrix A with the property that $\Delta_I(A) \geq 0$ and $K_I > 0$ for all $I \in \binom{[M]}{N}$, we have a positive definite τ-function which leads to a regular solution. Such matrix A is called a *totally nonnegative* matrix, and we will discuss the details of those matrices in Chap. 5. The positivity of K_I can be achieved by assuming the ordering,

$$\kappa_1 < \kappa_2 < \cdots < \kappa_M.$$

Let us write the phase function $\theta_j(x, y, t)$ of the exponential function $E_j^{(0)} = e^{\theta_j}$ in the form,

$$\theta_j(x, y, t) = p_j x + q_j y + \omega_j t.$$

Then the wave parameters $\mathbf{v}_j := (p_j, q_j)$ for the KP soliton is a point on the parabola, $q = p^2$, i.e. $p_j = \kappa_j$ and $q_j = \kappa_j^2$ (see Fig. 3.1).

Of particular interest is a *duality* that exists between the contour plot of a KP soliton and the triangulation of a polygon inscribed in the parabola. In Fig. 3.1, we present the pentagon inscribed in the parabola whose vertices are $\mathbf{v}_j = (\kappa_j, \kappa_j^2)$ for $j = 1, \ldots, 5$. Let us define the weighted vertex $\hat{\mathbf{v}}_j = (\kappa_j, \kappa_j^2, \kappa_j^3 t) \in \mathbb{R}^3$, where the third coordinate gives the weight $\kappa_j^3 t$. Then we consider the convex hull of those points in $\mathbb{R}^3$, which gives a convex polytope denoted by P. The projection of all upper faces of P onto the pq-plane gives a *regular* (or coherent) triangulation of the pentagon (see e.g. [32] for the general concept of triangulation). One can show that the triangulation given in Fig. 3.1 corresponds to the case for $t < 0$ (see Problem 3.3). Having a triangulation of the polygon, we can draw a *skeleton* graph of the contour plot of the solution $u(x, y, t)$ for fixed t corresponding to the triangulation. (A graph of this type will be called a *soliton graph*, and we will study those graphs in Chap. 8.) To construct the soliton graph, we first recall that line-soliton of $[i, j]$-type is localized along the line $L_{[i,j]} : \theta_i = \theta_j$, which is

$$(\hat{\mathbf{v}}_i - \hat{\mathbf{v}}_j) \cdot (x, y, t) = 0 \quad \text{i.e.} \quad x + (\kappa_i + \kappa_j)y + (\kappa_i^2 + \kappa_j^2 + \kappa_i \kappa_j)t = 0.$$

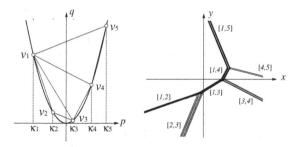

Fig. 3.1 A triangulated pentagon inscribed in a parabola $q = p^2$ and the contour plot of the KP soliton $u(x, y, -10)$. The parameters are $(\kappa_1, \ldots, \kappa_5) = (-2, -0.8, 0.4, 1.5, 2.4)$ and $A = (1\,1\,1\,1\,1)$. Here $\mathbf{v}_j = (\kappa_j, \kappa_j^2)$. Note a *duality* between the triangulation of the pentagon and the KP soliton

That is, the edge between the vertices $\{v_i, v_j\}$ is perpendicular to the line $L_{[i,j]}$ of the $[i, j]$-soliton. Then we start with the triangle with $\{v_1, v_4, v_5\}$ and construct the soliton graph of the corresponding Y-soliton with $[1, 5]$-, $[1, 4]$- and $[4, 5]$-solitons. The intersection point can be found from the resonant equation $\theta_1 = \theta_4 = \theta_5$. Now consider the triangle with $\{v_1, v_4, v_3\}$ and construct the soliton graph of the Y-soliton with $[1, 4]$-, $[1, 3]$- and $[3, 4]$-solitons, where $[1, 4]$-soliton is the same one constructed previously. Finally, consider the triangle with $\{v_1, v_2, v_3\}$ and do the same process. Then we get the soliton graph shown in Fig. 3.1.

Notice that the above construction of the soliton graph can be used for other solitons (see the following sections). We would like to invite the readers to see [57] for the details of the triangulations and the soliton graphs (see also Problem 8.4).

3.2.2   Two-Dimensional Toda Lattice Solitons

The two-dimensional Toda lattice equation has the following form with the variables $V_n = V_n(x, t)$ for $n \in \mathbb{Z}$ (see, e.g., [56, 128]),

$$\left(\frac{\partial^2}{\partial t^2} - \frac{\partial^2}{\partial x^2} \right) \ln(1 + V_n) = V_{n+1} - 2V_n + V_{n-1}.$$

Expressing V_n in terms of the $\tau^{(n)}$-function,

$$V_n(x, t) = -\left(\frac{\partial^2}{\partial t^2} - \frac{\partial^2}{\partial x^2} \right) \ln \tau^{(n)}(x, t),$$

we have the bilinear equation (3.4) for $\tau^{(n)}(x, y, t)$ with

$$x = x_1 - x_{-1}, \qquad t = x_1 + x_{-1}, \qquad n = s \in \mathbb{Z}.$$

We take the exponential function $E_j^{(n)}(x, t)$ as a simple solution of (3.6),

$$E_j^{(n)}(x, t) = \lambda_j^n \exp\left[\frac{1}{2} \left(\lambda_j - \frac{1}{\lambda_j} \right) x + \frac{1}{2} \left(\lambda_j + \frac{1}{\lambda_j} \right) t \right]$$

$$= (\pm 1)^n \exp\left(\pm x \sinh \phi_j \pm t \cosh \phi_j + n\,\phi_j \right) =: (\pm 1)^n e^{\theta_j^\pm(x,t)+n\phi_j}.$$

where $\lambda_j = \pm \exp \phi_j \in \mathbb{R}$. Note that writing the exponent in the form $\theta_j^\pm(x, t) := \pm(p_j x + q_j y)$, the parameters (p_j, q_j) are lying on the hyperbola, $q^2 - p^2 = 1$, and the variable ϕ_j represents the area of the region bounded by the hyperbola and the lines $q_j x = p_j y$ and $x = 0$ (the ϕ_j with negative x should be considered as $-$(the corresponding area)).

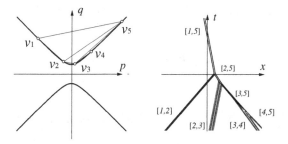

Fig. 3.2 A triangulated pentagon inscribed in a hyperbola $q^2 - p^2 = 1$ and the contour plot of $V^{(-10)}(x, t)$. The parameters are $(\phi_1, \ldots, \phi_5) = (-2, -0.8, 0.4, 1.5, 2.4)$ and $A = (1\,1\,1\,1\,1)$. Here $\mathbf{v}_j = (\sinh \phi_j, \cosh \phi_j)$. Note again the *duality* between the triangulation of the pentagon and the corresponding soliton solution

We parametrize the hyperbola as $(p, q) = (\sinh \phi, \pm \cosh \phi)$ where the sign gives a specific branch of the curve, i.e. $+(-)$ corresponds to the upper (lower) branch (see Fig. 3.2).

Then we have the Wronskian $E_I^{(n)} = \mathrm{Wr}(E_{i_1}^{(n)}, \ldots, E_{i_N}^{(n)}) = \Lambda_I^{(n)} e^{\Theta_I^{\pm}}$ with $\Theta_I^{\pm} = \sum_k \theta_{i_k}^{\pm}$ and

$$\Lambda_I^{(n)} = \left(\prod_{k=1}^N \lambda_{i_k}^n \right) \prod_{j>k} (\lambda_{i_j} - \lambda_{i_k}) = (\pm 1)^{nN} 2^{\frac{N(N-1)}{2}} e^{(n+\frac{N-1}{2}) \sum_k \phi_{i_k}} Sh_I(\phi)$$

with $Sh_I(\phi) = \prod_{j>k} \sinh \frac{1}{2} (\phi_{i_j} - \phi_{i_k})$. The positivity of $Sh_I(\phi)$ is obtained, if we order the $\phi_j \in \mathbb{R}$ as

$$\phi_1 < \phi_2 < \cdots < \phi_M.$$

Then the τ-functions for the two-dimensional Toda lattice equation are given by

$$\tau^{(n)}(x, t) = (\pm 1)^{nN} 2^{\frac{N(N-1)}{2}} \sum_{I \in \binom{[M]}{N}} e^{\frac{N-1}{2} \Phi_I} \Delta_I(A)\, Sh_I(\phi)\, e^{\Theta_I^{\pm}(x, t) + n\Phi_I},$$

where $\Phi_I = \sum_k \phi_{i_k}$ for the index set $I = \{i_1, \ldots, i_N\}$. Notice that the common factor $(\pm 1)^n N 2^{\frac{N(N-1)}{2}}$ in the τ-function does not contribute the solution $V_n(x, t)$. One can also see that the soliton becomes singular if we choose a pair (ϕ_i, ϕ_j) from the different branch of hyperbola.

The right panel in Fig. 3.2 illustrates the contour plot of the solution $V^{(n)}(x, t)$ on the xt-plane for $n = -10$. Note again a duality between the triangulation of the pentagon inscribed in the hyperbola and the contour plot of the solution (see Problem 3.3 for the duality).

Remark 3.1 A line-soliton of $[i, j]$-type of the two-dimensional Toda lattice equation has the form

$$V_n = A_{[i,j]} \text{sech}^2 \left[k_{[i,j]}(x + c_{[i,j]}t) + (\phi_i - \phi_j)n \right],$$

where

$$A_{[i,j]} = 4 \sinh^2 \tfrac{1}{2}(\phi_i - \phi_j), \quad k_{[i,j]} = \sqrt{A_{[i,j]}} \cosh \tfrac{1}{2}(\phi_i + \phi_j), \quad c_{[i,j]} = \tanh \tfrac{1}{2}(\phi_i + \phi_j).$$

Then, for each n, the solution V_n describes an interaction (*fusion*) pattern of several particles propagating with the speeds $c_{[i,j]}$ (in Fig. 3.2, there are five particles). Note that for fixed t, those particles behave in a similar manner along the lattice coordinate n.

3.2.3 The Davey-Stewartson (DS) Solitons

Here we consider the Davey-Stewartson equation of type II and defocusing case, which has the form (see e.g. [3, 36, 42, 52, 101] for some explicit soliton solutions),

$$iq_t + q_{xx} - q_{yy} + 4qQ + 2|q|^2 q = 0,$$
$$Q_{xx} + Q_{yy} = -(|q|^2)_{xx}.$$

If we assume the complex function $q(x, y, t)$ and the real function $Q(x, y, t)$ in the forms

$$q = \frac{\tau^{(s+1)}}{\tau^{(s)}} e^{2it}, \quad \text{and} \quad Q = \frac{\partial^2}{\partial x^2} \ln \tau^{(s)}.$$

then these τ-functions, $\tau^{(s)}$ and $\tau^{(s\pm1)}$ with $\tau^{(s+1)} = [\tau^{(s-1)}]^*$, the complex conjugate of $\tau^{(s-1)}$, satisfy the bilinear equations (3.2), (3.3) and (3.4), where the variables (x, y, t) are expressed by the following formulas with $(x_{-k}, x_k = x_{-k}^*)$ for $k = 1, 2$:

$$x = i(x_1 - x_{-1}), \quad y = x_1 + x_{-1}, \quad t = i(x_2 - x_{-2}).$$

We take the following exponential function as a solution of (3.6):

$$E_j^{(s)}(x, y, t) = \lambda_j^s \exp\left[\frac{1}{2i}\left(\lambda_j - \frac{1}{\lambda_j}\right)x + \frac{1}{2}\left(\lambda_j + \frac{1}{\lambda_j}\right)y + \frac{1}{2i}\left(\lambda_j^2 - \frac{1}{\lambda_j^2}\right)t \right]$$

$$= \exp\left(-x \sin \psi_j + y \cos \psi_j - t \sin(2\psi_j) - is\psi_j\right) =: e^{\theta_j(x,y,t) - is\psi_j},$$

where $\lambda_j = e^{-i\psi_j} \in \mathbb{C}$ with $0 \le \psi_j < 2\pi$. Notice that $E_j^{(s)}$ is a complex factor $e^{-is\psi_j}$. In this setting, the parameter (p_i, q_i) in $\theta_i(x, y, t) := p_i x + q_i y + \omega_i t$ is

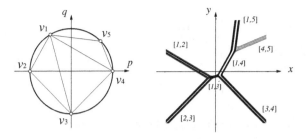

Fig. 3.3 A pentagon inscribed in a unit circle and the contour plot of $Q(x, y, -10)$. The parameters are $(\psi_1, \ldots, \psi_5) = (\pi/6, \pi/2, \pi, 3\pi/2, 7\pi/4)$ and $A = (1\,1\,1\,1\,1)$. Here $\mathbf{v}_i = (-\sin\psi_i, \cos\psi_i)$. Also note the *duality* between the triangulation of the pentagon and the DS soliton

taken from a unit circle $(p, q) = (-\sin\psi, \cos\psi)$ where ψ is measured in the counterclockwise direction starting from the positive y-axis (see Fig. 3.3). With these $E_j^{(s)}$, we calculate the Wronskian $E_I^{(s)} = \mathrm{Wr}(E_{i_1}^{(s)}, \ldots, E_{i_N}^{(s)}) = \Lambda_I^{(s)} e^{\Theta_I}$ where $\Lambda_I^{(s)}$ is given by

$$\Lambda_I^{(s)} = \left(\prod_{k=1}^{N} \lambda_{i_k}^s\right) \prod_{j>k}(\lambda_{i_j} - \lambda_{i_k}) = (-2i)^{\frac{N(N-1)}{2}} e^{-i\left(s+\frac{N-1}{2}\right)\sum_k \psi_{i_k}} S_I(\psi),$$

where $S_I(\psi) := \prod_{j>k} \sin\frac{1}{2}(\psi_{i_j} - \psi_{i_k})$. Choosing $s = -\frac{N-1}{2}$, we have

$$E_I^{\left(\frac{1-N}{2}\right)}(x, y, t) = (-2i)^{\frac{N(N-1)}{2}} S_I(\psi) e^{\Theta_I(x,y,t)}.$$

Then the τ-function is given by

$$\tau^{\left(\frac{1-N}{2}\right)}(x, y, t) = (-2i)^{\frac{N(N-1)}{2}} \sum_{I \in \binom{[M]}{N}} \Delta_I(A) S_I(\psi) e^{\Theta_I(x,y,t)}.$$

Note again that the factor $(-2i)^{\frac{N(N-1)}{2}}$ does not give any contribution to the solution. The positivity of $S_I(\psi)$ can be obtained by taking the ordering

$$0 \le \psi_1 < \psi_2 < \cdots < \psi_M < 2\pi.$$

Then choosing a matrix A whose minors are all nonnegative, one can see that the τ-function ignoring the factor $(-2i)^{\frac{N(N-1)}{2}}$ is positive-definite, and we obtain the regular solutions $q(x, y, t)$ and $Q(x, y, t)$ for all time $t \in \mathbb{R}$.

Example 3.1 Consider the case with $M = 4$ and $N = 1$. Then the τ-function is given by

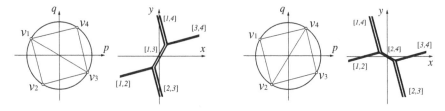

Fig. 3.4 Triangulations of 4-gon and the contour plots of DS soliton

$$\tau^{(0)} = \sum_{j=1}^{4} a_i e^{\theta_j} \quad \text{with} \quad \theta_j = p_j x + q_j + \omega_j t,$$

where $(p_j, q_j, \omega_j) = (-\sin \psi_j, \cos \psi_j, -\sin 2\psi_j)$. Figure 3.4 shows the contour plots of the DS soliton with ψ-parameters are $(\frac{\pi}{3}, \frac{5\pi}{6}, \frac{4\pi}{3}, \frac{11\pi}{6})$. The left (right) set of two figures shows the triangulation of the square with the diagonal edge connecting $\{v_1, v_3\}$ ($\{v_2, v_4\}$) and the corresponding soliton solution for $t < 0$ ($t > 0$). Each $[i, j]$ shows the line-soliton of $[i, j]$-type. In [99], this type of phenomena is called *soliton reconnection*.

Problems

3.1 Derive the bilinear equations (3.2), (3.3) and (3.4) from the bilinear identity.

3.2 Give the explicit form of one-soliton solution for the Davey-Stewartson equations. In particular, note that the soliton solution $Q(x, y, t)$ of $[i, j]$-type vanishes if $\sin \psi_i = \sin \psi_j$, i.e. there is no DS soliton parallel to the x-axis.

3.3 Consider a set of four points $\{\hat{v}_j \in \mathbb{R}^3 : j = 1, \ldots, 4\}$ where each point $\hat{v}_j = (p_j, q_j, \omega_j)$ is on a conic curve. Also assume that those points are ordered as explained in this chapter (i.e. counterclockwise direction). Let D be a determinant defined by

$$D := \begin{vmatrix} 1 & p_1 & q_1 & \omega_1 \\ 1 & p_2 & q_2 & \omega_2 \\ 1 & p_3 & q_3 & \omega_3 \\ 1 & p_4 & q_4 & \omega_4 \end{vmatrix}.$$

Show that if $D \neq 0$, the convex hull of this point set is a tetrahedron denoted by P. The projections of all upper faces of P form a *regular* triangulation of the quadrilateral whose vertices are $v_j = (p_j, q_j)$. Then prove the following lemma:

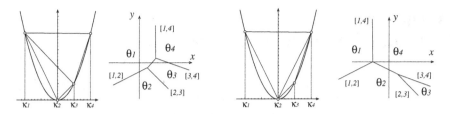

Fig. 3.5 Triangulations and the soliton graphs. The left two figures shows the triangulation with the diagonal connecting 1 and 3 vertices and the corresponding soliton graph for $t < 0$. The right two figures are those for $t > 0$. The κ-parameters are $(-2, 0, 1, 2)$. Each θ_i shows the dominant plane in that region

Lemma 3.1 *If $D > 0$ (resp. $D < 0$), the projection of upper face of* P *gives the triangulation with the diagonal connecting the vertices $\{v_2, v_4\}$ (resp. $\{v_1, v_3\}$).*

Using this lemma, show that the triangulations of pentagons in Figs. 3.1, 3.2 and 3.3 are all regular.

3.4 Following the arguments given in the Subsect. 3.2.1, construct the soliton graphs shown in Fig. 3.5, which are *dual* to the triangulations of the quadrilateral obtained by Lemma 3.1. Here, the duality means a bijection between a vector $\hat{v} = (p, q, \omega(t))$ and a line $L : px + qy + \omega(t) = 0$. Then the edge in the triangulation connecting the vertices $\{v_i, v_j\}$ corresponds to the line-soliton of $[i, j]$-type.

Chapter 4
Introduction to the Real Grassmannian

Abstract This chapter gives a brief introduction to the real Grassmannian $\mathrm{Gr}(N, M)$, the set of N-dimensional subspaces in $\mathbb{R}^M$, which provides a foundation of a classification of the KP solitons. A point of $\mathrm{Gr}(N, M)$ can be represented by an $N \times M$ matrix of full rank. We introduce the Schubert decomposition of $\mathrm{Gr}(N, M)$ and label each Schubert cell using a Young diagram and a permutation in the symmetric group S_M. We also introduce a combinatorial tool called the *pipedream* over the Young diagram, which gives a graphical interpretation of the permutation [103]. The pipedream will be useful to describe the spatial structure of the KP soliton as we will see in the later chapters. (See, for example, [15, 44, 45, 49] for the general information on the Grassmannian, the Young diagram and the symmetric group of permutations.)

4.1 Soliton Solutions and the Grassmannians

In Chap. 1, we presented the KP solitons in the form $u(x, y, t) = 2\partial_x^2 \ln \tau(x, y, t)$, where the τ-function is given by the Wronskian determinant,

$$\tau(x, y, t) = \mathrm{Wr}(f_1, f_2, \ldots, f_N). \tag{4.1}$$

The set of functions $\{f_i(x, y, t) : i = 1, \ldots, N\}$ are expressed by

$$(f_1, f_2, \ldots, f_N) = (E_1, E_2, \ldots, E_N) A^T, \tag{4.2}$$

where A is a full rank $N \times M$ matrix, A^T is the transpose of A, and

$$E_j(x, y, t) = e^{\theta_j(x,y,t)} \quad \text{with} \quad \theta_j(x, y, t) = \kappa_j x + \kappa_j^2 y + \kappa_j^3 t.$$

Thus, each soliton solution is determined by the κ-parameters and the matrix A. In this chapter, we describe matrix A as a point of the real Grassmannian $\mathrm{Gr}(N, M)$ and a classification of those matrices in terms of several decompositions of $\mathrm{Gr}(N, M)$. In particular, we discuss $N \times M$ matrices A, whose minors are all non-negative, and how they play a key roll for the *regularity* of the KP solitons.

© The Author(s) 2017

Y. Kodama, *KP Solitons and the Grassmannians*,

SpringerBriefs in Mathematical Physics 22, DOI 10.1007/978-981-10-4094-8_4

4.2 The Real Grassmannian

The real Grassmannian $\text{Gr}(N, M)$ is the set of all N-dimensional subspaces of $\mathbb{R}^M$. A point ξ of $\text{Gr}(N, M)$ can be expressed by an N-frame of vectors, $\xi_i \in \mathbb{R}^M$ for $i = 1, \ldots, M$,

$$\xi = [\xi_1, \xi_2, \ldots, \xi_N] \quad \text{with} \quad \xi_i = \sum_{j=1}^{M} a_{i,j} e_j \in \mathbb{R}^M,$$

where $\{e_i : i = 1, \ldots, M\}$ is the standard basis of $\mathbb{R}^M$, and $(a_{i,j}) =: A$ is a full rank $N \times M$ matrix. Then $\text{Gr}(N, M)$ can be embedded into the projectivization of the exterior product space, denoted by $\mathbb{P}(\wedge^N \mathbb{R}^M)$, which is called the *Plücker embedding*,

$$\text{Gr}(N, M) \hookrightarrow \mathbb{P}(\wedge^N \mathbb{R}^M)$$
$$[\xi_1, \ldots, \xi_N] \mapsto \xi_1 \wedge \cdots \wedge \xi_N.$$

Here the element of $\wedge^N \mathbb{R}^M$ is expressed as

$$\xi_1 \wedge \cdots \wedge \xi_N = \sum_{1 \le i_1 < \cdots < i_N \le M} \Delta_{i_1, \ldots, i_N}(A) \, e_{i_1} \wedge \cdots \wedge e_{i_N},$$

where the coefficients $\Delta_{i_1, \ldots, i_N}(A)$ are $N \times N$ minors of the matrix A called the *Plücker coordinates*. (Notice that those coordinates already appear in Chap. 1 as the coefficients of the exponential terms in the τ-functions). Using the usual inner product on $\wedge^N \mathbb{R}^M$, $\langle \cdot, \cdot \rangle \to \mathbb{R}$, we have

$$\Delta_{i_1, \ldots, i_N}(A) = \langle e_{i_1} \wedge \cdots \wedge e_{i_N}, \xi_1 \wedge \cdots \wedge \xi_N \rangle.$$

In this book, we often identify the point $\xi \in \text{Gr}(N, M)$ as a full-rank $N \times N$ matrix A modulo left multiplication by nonsingular $N \times N$ matrices. In other words, two $N \times M$ matrices represent the same point in $\text{Gr}(N, M)$ if and only if they can be obtained from each other by row operations. That is, for any $H \in \text{GL}_N(\mathbb{R})$, the new set $\{\eta_i : i = 1, \ldots, N\}$ given by

$$[\eta_1, \ldots, \eta_N] = [\xi_1, \ldots, \xi_N] \cdot H$$

spans the same N-dimensional subspace. Since $H \in \text{GL}_N$ gives a row operation to the A-matrix, choosing appropriate H, one can put $H^T A$ in a canonical form (H^T is the transpose of H) called a *reduced row echelon form* (RREF). We then have the isomorphism

$$\text{Gr}(N, M) \cong \text{GL}_N(\mathbb{R}) \setminus M_{N \times M}(\mathbb{R}),$$

where $M_{N \times M}(\mathbb{R})$ is the set of all $N \times M$ full-rank matrices. This then gives the dimension of $\text{Gr}(N, M)$ as $\dim \text{Gr}(N, M) = MN - N^2 = N(M - N)$.

Definition 4.1 Recall that $\binom{[M]}{N}$ denotes the set of all N-element subsets of $[M] :=$ $\{1, \ldots, M\}$. We consider each element $I \in \binom{[M]}{N}$ an ordered set, $I = \{i_1 < \cdots < i_N\}$. We then define the *matroid* associated with the matrix A as

$$\mathscr{M}(A) := \left\{ I \in \binom{[M]}{N} : \Delta_I(A) \neq 0 \right\}.$$

Notice that if A represents a *generic* point of $\mathrm{Gr}(N, M)$, then $\mathscr{M}(A) = \binom{[M]}{N}$, i.e. the Plücker coordinates are all nonzero.

Remark 4.1 In combinatorics, a matroid of rank N on the set $[M]$ is defined as a nonempty collection $\mathscr{M}$ of N-element subsets in $[M]$ that satisfies the *exchange axiom*: For any distinct elements $I, J \in \mathscr{M}$ and $i \in I \setminus J$, there exists $j \in J \setminus I$ such that $(I \setminus \{i\}) \cup \{j\} \in \mathscr{M}$.

Definition 4.2 It is known that the matroid $\mathscr{M}(A)$ is a partially ordered set with the *lexicographical order*, which is defined as follows: For $I = \{i_1 < i_2 < \cdots < i_N\}$ and $J = \{j_1 < j_2 < \cdots < j_N\}$, we say I is lexicographically smaller (resp. larger) than J, denoted by $I < J$ (resp. $I > J$), if $i_k < j_k$ (resp. $i_k > j_k$) for the first k where i_k and j_k differ. There exist the lexicographically maximal and minimal elements in $\mathscr{M}(A)$, and in particular, the lexicographically minimal element is the index set of (*left*) pivots of the matrix A in RREF.

Remark 4.2 In standard notation, the RREF for a matrix A has an *upper* triangular structure, i.e. all the entries *before* each pivot element are zero. One can also have a different RREF for the same matrix, which has a *lower* triangular structure, i.e. all the entries *after* each pivot element are zero. Then the index set of the later pivots is the lexicographically maximal elements. We may refer to the former as the *left* pivot (or just pivot), and the latter as the *right* pivot (see Chap. 8).

Example 4.1 A *generic* point in $\mathrm{Gr}(N, M)$ can be expressed as the following matrix in the RREF:

$$\begin{pmatrix} 1 & \cdots & 0 & * & \cdots & * \\ \vdots & \ddots & \vdots & \vdots & \ddots & \vdots \\ 0 & \cdots & 1 & * & \cdots & * \end{pmatrix} \quad \text{or} \quad \begin{pmatrix} * & \cdots & * & 1 & \cdots & 0 \\ \vdots & \ddots & \vdots & \vdots & \ddots & \vdots \\ * & \cdots & * & 0 & \cdots & 1 \end{pmatrix},$$

where the 1's are called the *pivot ones*, i.e. the index set of the (left) pivots is given by the lexicographically minimal set $I = \{1, 2, \ldots, N\}$ and all the minors are nonzero. The lexicographically maximal set is $J = \{M - N + 1, \ldots, M\}$, which is the set of the right pivots of the RREF given at the right.

4.3 The Schubert Decomposition

The Schubert decomposition of $\mathrm{Gr}(N, M)$ is defined by

$$\mathrm{Gr}(N, M) = \bigsqcup_{I \in \binom{[M]}{N}} \Omega_I,$$

where Ω_I is defined by the set of all matrices whose pivots are given by $I = \{i_1, \ldots, i_N\}$. It is known that each Ω_I has a cellular structure, and it is called a *Schubert cell* associated to the index set I. The Schubert cell Ω_I has the dimension

$$\dim(\Omega_I) = N(M - N) - \sum_{k=1}^{N} (i_k - k).$$

In the next subsection, we introduce the *Young diagram* to express the index set I for an alternative parametrization of the Schubert cells.

Example 4.2 The element of $\Omega_{\{1,4,5\}} \subset \mathrm{Gr}(3, 7)$ has the RREF

$$A = \begin{pmatrix} 1 * * 0\,0 * * \\ 1\,0 * * \\ 1 * * \end{pmatrix},$$

whose dimension is $\dim(\Omega_{\{1,4,5\}}) = 3 \times 4 - (0 + 2 + 2) = 8$, which is the number of free parameters indicated by "$*$" in A in the RREF.

4.3.1 Young Diagrams for a Parametrization of the Schubert Cells

For each pivot set $I = \{i_1, \ldots, i_N\}$, we consider a lattice path starting from the top right corner and ending at the bottom left corner with the label $\{1, \ldots, M\}$, so that the pivot indices appear at the vertical paths as shown in the diagram below.

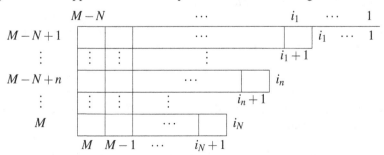

(Note that the labeling of the boundary of the Young diagram is different from the previous definition in Sect. 1.2.) For the case with the pivot set $I = \{1, \ldots, N\}$, we write $Y = (M - N)^N$ representing an $N \times (M - N)$ rectangular diagram, which is the top cell of $\mathrm{Gr}(N, M)$, i.e. the largest dimensional cell. Listing the number of boxes in each row gives a partition $\lambda \subset (M - N)^N$ of the total number of the boxes of the diagram. We recall that the partitions λ are in bijection with N-element subset $I \subset [M]$, i.e. we have $\lambda_1 \geq \lambda_2 \geq \cdots \geq \lambda_N$ with

$$\lambda_k = M - N - (i_k - k).$$

We then identify $\lambda = (\lambda_1, \ldots, \lambda_N)$ as the index set $I = (i_1, \ldots, i_N)$ and sometimes write $I = I_\lambda$.

Definition 4.3 The Young diagram $\lambda = (\lambda_1, \ldots, \lambda_N)$ is *irreducible* if

(a) the first row has the maximum number of boxes, i.e. $\lambda_1 = M - N$ (or simply $i_1 = 1$), and
(b) the last row has at least one box, i.e. $\lambda_N \geq 1$ (or simply $i_N < M$).

The dimension of the cell Ω_I is given by the number of boxes denoted by $|\lambda|$ in the corresponding Young diagram λ, i.e.

$$\dim(\Omega_I) = |\lambda| = \sum_{k=1}^{N}[M - N - (i_k - k)] = N(M - N) - \sum_{k=1}^{N}(i_k - k).$$

Then the Schubert decomposition can also be expressed using the Young diagrams,

$$\mathrm{Gr}(N, M) = \bigsqcup_{\lambda \subset (M-N)^N} \Omega_\lambda \quad \text{with} \quad \dim(\Omega_\lambda) = |\lambda|.$$

Note here that the Schubert cell Ω_λ can be written as

$$\Omega_\lambda = \{A \in \mathrm{Gr}(N, M) : I_\lambda \text{ is the lexicographically minimal base of } \mathcal{M}(A)\}.$$

We now recall that there is also a bijection between the set of partitions $\{\lambda \subset (M - N)^N\}$ and a subset $S_M^{(N)}$ of S_M defined below.

Definition 4.4 ([15]) Let S_M be the symmetric group of permutations on $[M]$ and let s_j be the simple reflection $s_j : j \leftrightarrow j + 1$ which satisfies

$$s_i s_j = s_j s_i, \quad \text{if } |i - j| \geq 2, \quad \text{and} \quad s_i s_{i+1} s_i = s_{i+1} s_i s_{i+1}.$$

S_M is then generated by the set of the simple reflections, denoted by $S_M = \langle s_1, \ldots, s_{M-1} \rangle$. We define a subset of S_M as

$$S_M^{(N)} := \{\text{the set of minimum length representatives of } S_M/P_N\},$$

where P_N is a maximum parabolic subgroup of S_M generated by the elements $\{s_1, \ldots, s_{M-1}\}$ without s_{M-N},

$$P_N = \langle s_1, \ldots, \widehat{s_{M-N}}, \ldots, s_{M-1} \rangle \cong S_{M-N} \times S_N.$$

Note that $S_M^{(N)}$ can also be defined by

$$S_M^{(N)} = \{\sigma \in S_M : \ell(\sigma s) \geq \ell(\sigma), \forall s \in P_N\}.$$

Here $\ell(\sigma)$ represents the *length* of $\sigma \in S_M$ in a *reduced* expression, i.e. $\sigma = s_{j_1} s_{j_2} \cdots s_{j_p}$ is a reduced expression if p is minimal. Note that any element $\sigma \in S_M^{(N)}$ must end with the letter $s_{M-N} \notin P_N$.

The set $S_M^{(N)}$ is a partially ordered set (*poset*) with the *Bruhat order* in S_M. Here, the Bruhat order is defined as follows: Let (v, w) be a pair of elements in S_M. We say that v and w are in the Bruhat order, denoted by $v \leq w$, if any reduced expression for w contains a subexpression that is a reduced expression for v.

Now we express a Young diagram $\lambda \subset (M - N)^N$ by a reduced expression $w \in S_M^{(N)}$. First we place s_j in each box in the $N \times (M - N)$ rectangular Young diagram corresponding to the top cell as shown below.

This labeling shows the ordering used in the construction of the Young diagram starting from the zero dimensional cell, which corresponds to the empty diagram. The pivot set of the element in this cell is given by $I_\emptyset = \{M - N + 1, M - N + 2, \ldots, M\}$. Applying the simple reflection s_{M-N} to $I_\emptyset$, one can obtain a new diagram corresponding to the one-dimensional cell whose pivot set is given by $I_\square = \{M - N, M - N + 2, \ldots, M\}$. Then applying either s_{M-N-1} or s_{M-N+1}, we obtain a new diagram having two boxes which corresponds to a two-dimensional cell. Here the new box was added either at the right or the bottom side of the previous box. Then any Young subdiagram $\lambda \subset (M - N)^N$ can be obtained by applying an *appropriate* generator s_j in each step, which gives a new box labeled s_j. Note here that one can create a box if and only if the new pivot set obtained by $s_j \cdot \{i_1, \ldots, i_N\}$ is different from the old set $\{i_1, \ldots, i_N\}$. We denote this as

$$\lambda \xrightarrow{s_j} \lambda' \quad \text{iff} \quad \lambda' \neq \lambda,$$

where we identify the Young diagram λ with the pivot set $I_\lambda = \{i_1, \ldots, i_N\}$. One should also note that the ordering in the construction of a particular diagram is not unique, which reflects the non-uniqueness of the *reduced* expressions of the element in S_M (recall $s_i s_j = s_j s_i$ if $|i - j| > 1$). For example, we have the following construction scheme for $\mathrm{Gr}(2, 5)$. We start with the pivot set $I_\emptyset = \{4, 5\}$. Then apply s_3 to generate a cell with $I_\square = \{3, 5\}$. By applying simple reflections s_j, we get all the elements in $\binom{[5]}{2}$,

$$
\begin{array}{ccccccc}
\{4, 5\} & \xrightarrow{s_3} & \{3, 5\} & \xrightarrow{s_2} & \{2, 5\} & \xrightarrow{s_1} & \{1, 5\} \\
 & & \scriptstyle s_4 \downarrow & & \scriptstyle s_4 \downarrow & & \scriptstyle s_4 \downarrow \\
 & & \{3, 4\} & \xrightarrow{s_2} & \{2, 4\} & \xrightarrow{s_1} & \{1, 4\} \\
 & & & & \scriptstyle s_3 \downarrow & & \scriptstyle s_3 \downarrow \\
 & & & & \{2, 3\} & \xrightarrow{s_1} & \{1, 3\} \\
 & & & & & & \scriptstyle s_2 \downarrow \\
 & & & & & & \{1, 2\}
\end{array}
$$

The corresponding Young diagrams with the generators s_j are given by the following diagram,

One should note that the diagram is commutative, i.e. we have the following:

Lemma 4.1 *Let $\{\lambda_1, \ldots, \lambda_4\}$ be the set of the Young diagrams generated by the simple reflections s_i and s_j such that*

$$
\begin{array}{ccc}
\lambda_1 & \xrightarrow{s_i} & \lambda_2 \\
\scriptstyle s_j \downarrow & & \scriptstyle s_j \downarrow \\
\lambda_3 & \xrightarrow{s_i} & \lambda_4
\end{array}
$$

Then we have $|i - j| \geq 2$, such that $s_i s_j = s_j s_i$.

This Lemma implies that $S_M^{(N)}$ consists of *fully* commutative elements, i.e. any element $w \in S_M^{(N)}$ if every pair of reduced expressions for w is related by a sequence of relations of the form $s_i s_j = s_j s_i$.

Then it follows from [119] that a reduced expression $w \in S_M^{(N)}$ corresponds to a certain *reading order* of the boxes of the Young $\lambda \subset (M - N)^N$. Specifically, let Q^N be the poset whose elements are the boxes in an $N \times (M - N)$ rectangle with each box labeled by a simple reflection as explained above (see also the left diagram in the figure below). If b_1 and b_2 are two adjacent boxes such that b_2 is immediately to the left or immediately above b_1, we have a cover relation $b_1 \lessdot b_2$ in Q^N. The partial order on Q^N is the transitive closure of $\lessdot$. Let $w_0 \in S_M^{(N)}$ denote the longest element in $S_M^{(N)}$. Then the set of reduced expressions of w_0 can be obtained by choosing a linear extension of Q^N and writing down the corresponding expression in the simple reflections s_i's. We call such a linear extension a *reading order* as shown in the figure below. The middle and right diagrams in the figure below show two examples of a *reading order* of the poset Q^N, where $N = 3$ and $M = 8$.

s_5	s_4	s_3	s_2	s_1
s_6	s_5	s_4	s_3	s_2
s_7	s_6	s_5	s_4	s_3

15	14	13	12	11
10	9	8	7	6
5	4	3	2	1

15	12	9	6	3
14	11	8	5	2
13	10	7	4	1

For each Young diagram $\lambda \subset (M - N)^N$, a reading order for λ can be expressed by filling the numbers from 1 to $|\lambda| = \dim(\Omega_\lambda)$, such that the entries decrease from left to right in rows and increase from top to bottom in columns. The corresponding word is then a reduced expression of the permutation $w \in S_M^{(N)}$. Moreover, all reduced expressions of w can be obtained by varying the reading order. For example, consider the Young diagram $\lambda = $ ⊞ in Gr(2, 5) corresponding to the pivot set $I_\lambda = \{1, 3\}$ (see the example in the previous page). We first choose a reading order in the Young diagram. The middle and right diagrams below show two examples of a reading order. Then a reduced word of w can be expressed by $w = s_3 s_4 s_1 s_2 s_3$ from the middle diagram or $w = s_1 s_3 s_2 s_4 s_3$ from the right diagram:

s_3	s_2	s_1
s_4	s_3	

5	4	3
2	1	

5	3	1
4	2	

Recall that each reading order for the Young diagram λ is just the construction order of λ from the empty diagram where the newest box is labeled by the number 1 and the oldest box with s_{M-N} is labeled by the number $|\lambda|$ (see the transition diagram for Gr(2, 5) in the previous page).

For each reduced expression $w \in S_M^{(N)}$ and the corresponding Young diagram λ_w, the Schubert decomposition can be also expressed by

$$\text{Gr}(N, M) = \bigsqcup_{w \in S_M^{(N)}} \Omega_w \quad \text{with} \quad \dim(\Omega_w) = \ell(w) = |\lambda_w|. \tag{4.3}$$

Example 4.3 Consider the Schubert cells for Gr(3, 7).

The index sets corresponding to these diagrams are $\{1, 2, 3\}$, $\{1, 4, 5\}$, $\{2, 4, 6\}$, $\{3, 4, 7\}$, $\{1, 6, 7\}$ from left to right. The left diagram labels the top cell, and the corresponding permutation can be expressed as $w = s_3 s_4 s_5 s_6 s_2 s_3 s_4 s_5 s_1 s_2 s_3 s_4$. The middle one with $I = \{2, 4, 6\}$ is labeled by $w = s_6 s_4 s_5 s_2 s_3 s_4$.

Remark 4.3 It is known that the generating function of the number β_j of cells in jth dimension is given by the q-deformation of the binomial coefficient $\binom{M}{N}$,

$$p(q) = \sum_{\lambda \subset (M-N)^N} q^{|\lambda|} = \sum_{j=0}^{N(M-N)} \beta_j q^j = \begin{bmatrix} M \\ N \end{bmatrix}_q := \frac{[M]_q!}{[N]_q![M-N]_q!}$$

where $[k]_q! := [k]_q[k-1]_q \cdots [1]_q$ and $[k]_q$ is the q-analog of k,

$$[k]_q = \frac{1-q^k}{1-q} = 1 + q + \cdots + q^{k-1}.$$

It is also known that $p(q) = |\mathrm{Gr}(N, M; \mathbb{F}_q)|$, i.e. $p(q)$ gives the total number of points in $\mathrm{Gr}(N, M; \mathbb{C})$ over the finite field $\mathbb{F}_q$ with q elements (see e.g. [11]). For example, in the case of $\mathrm{Gr}(3, 7)$, we have

$$p(q) = \begin{bmatrix} 7 \\ 3 \end{bmatrix}_q = \frac{[7]_q[6]_q[5]_q}{[3]_q[2]_q[1]_q} = \frac{(1-q^7)(1-q^6)(1-q^5)}{(1-q^3)(1-q^2)(1-q)}$$
$$= 1 + q + 2q^2 + 3q^3 + 4q^4 + 4q^5 + 5q^6$$
$$+ 4q^7 + 4q^8 + 3q^9 + 2q^{10} + q^{11} + q^{12}.$$

Thus, for example, there are four five-dimensional Schubert cells in the decomposition, each of which is labeled by a Young diagrams with five boxes.

4.3.2 Pipedreams and Permutations

Let us define $\pi = w^{-1}$, such that the pivot indices are given by the excedances of π, i.e. if the index i is a pivot index, then $\pi(i) > i$. For example, the cell with the pivot set $I = \{1, 4, 5\}$ of $\mathrm{Gr}(3, 7)$ is labeled by $w = s_5 s_6 s_4 s_5 s_1 s_2 s_3 s_4$ and we have

$$\pi = \begin{pmatrix} 1\,2\,3\,4\,5\,6\,7 \\ 5\,1\,2\,6\,7\,3\,4 \end{pmatrix}.$$

In terms of the Young diagram, $w^{-1}\{1, 2, \ldots, M\}$ is computed as shown in the diagram below.

	4	3	2	1	
5	s_4	s_3	s_2	s_1	1
6	s_5	s_4	4 3		2
7	s_6	s_5	5		
	7	6			

Here there are two labelings from 1 to M starting from the top right (northeast) corner and ending at the bottom left (southwest) corner along the boundary of the

Young diagram: One labeling is to follow the path in the clockwise, and other is in the counterclockwise direction. Then each pair $(i, \pi(i))$ of indices appears at the opposite sides, where "i" is the label in the clockwise path, and "$\pi(i)$" is the label in the counterclockwise path. In the example above, we have $(1, 5), (2, 1), (3, 2)$ etc.

We now introduce a game called the *pipedream*, a graphical representation of the permutation $\pi = w^{-1}$ ([103], also see Sect. 5.3.2), which will be useful to compute the permutation when we have a refinement of the Schubert cell.

The pipedream for a Young diagram is constructed as follows: Staring at each edge of the southeast (*input*) boundary of the Young diagram, first we draw a path, which we call a *pipe*, and label it with the index of the edge. This gives the *first* index of the pair for each segment. Now we add a *bridge* between the pipes at each box and place a *white* and *black* vertex at the connection points as shown in the figure below. Then we make a path starting from the input boundary so that it turns right at a black vertex and left at a white vertex (these rules are called the *rules of the road*, [103]), and label this path with the index at the input boundary. This gives a *second* index for each segment which is just the index at the input boundary, such that each path now has a pair of indices. For example, below, the first indices in italics are written on the left side of the pipes and the permutation is then given by $w^{-1}(1, 2, \ldots, 7) = (5, 1, 2, 6, 7, 3, 4)$.

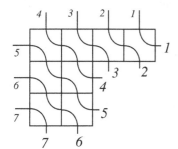 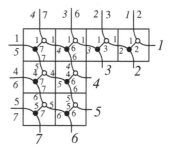

Notice that each bridge in a box located at the northwest corner of the Young diagram after removing the topmost row and the leftmost column has the pair (i, j) where i is a pivot index appearing at the east edge of the Young diagram and j is a non-pivot index appearing at the south edge of the Young diagram (see below).

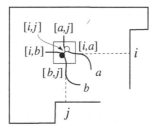

The pair of indices at each pipe-segment acts as a transposition, denoted by $[i, j]$, on the index sets of the Plücker coordinates. Assign the lexicographically minimum element in the matroid $\mathcal{M}(A)$ at the southeast boundary of the diagram. Then each pipe-segment gives the transposition on the index set. In this way, each region bounded by the pipes is labeled with a unique index set as shown in the right panels below.

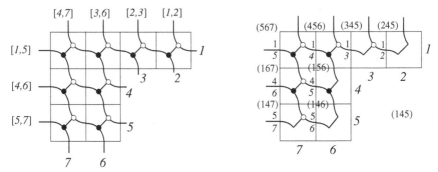

In the right figure above, we eliminate the edges having only one index and then eliminate the vertices of degree two. We call this diagram the *reduced pipedream*, and as will be shown in Chap. 8, it has an intimate connection with the soliton graph (see also Remark 4.4 below). In particular, the N-index subset of these indices appearing in the diagram forms a *basis* of the matroid $\mathcal{M}(A)$ denoted by $\mathcal{B}(A)$ because all the other elements in $\mathcal{M}(A)$ can be calculated from $\mathcal{B}(A)$. (We prove this more general case in the next chapter, see also [121]). In the example above, there are 22 nonzero minors and we have

$$\mathcal{M}(A) = \{\mathbf{145}, \mathbf{146}, \mathbf{147}, \mathbf{156}, 157, \mathbf{167}, \mathbf{245}, 246, 247, 256, 257, 267,$$
$$\mathbf{345}, 346, 347, 356, 357, 367, \mathbf{456}, 457, 467, \mathbf{567}\}.$$

Here the elements in bold-face are the indices appearing in the pipedream diagram. The set of those indices forms a basis of $\mathcal{M}(A)$, i.e.

$$\mathcal{B}(A) = \{145, 146, 147, 156, 167, 245, 345, 456, 567\},$$

and all other elements can be obtained from the Plücker relations. For example, the element Δ_{157} is given by

$$\Delta_{157}\Delta_{146} = \Delta_{145}\Delta_{167} + \Delta_{147}\Delta_{156}.$$

Also we have

$$\Delta_{246}\Delta_{145} = \Delta_{124}\Delta_{456} + \Delta_{146}\Delta_{245}.$$

Note here that $\Delta_{124} = 0$, i.e. the last relation is a two term Plücker relation.

Remark 4.4 There is an interesting connection between the soliton counter plots and the *reduced* pipedreams in Young diagrams. This will be discussed in Chap. 8. Here we remark on the cases of $\mathrm{Gr}(1, 3)$ and $\mathrm{Gr}(2, 3)$. The soliton solutions are given in Figs. 1.3 and 1.5. The reduced pipedreams corresponding to these solutions are illustrated in the figures below.

A trivalent vertex in the pipedream represents the interaction point of three line-solitons. The pipedream with a white vertex represents Y-soliton from $\mathrm{Gr}(1, 3)$, which has $[1, 3]$-soliton in $y \gg 0$ and $[1, 2]$- and $[2, 3]$-solitons in $y \ll 0$, and the pipedream with a black vertex gives Y-soliton from $\mathrm{Gr}(2, 3)$. This correspondence is the main motivation for connecting the soliton graphs and the Young diagrams with pipedreams.

Problems

4.1 Find the Schubert decomposition of $\mathrm{Gr}(3, 6)$ in the form

$$\mathrm{Gr}(3, 6) = \bigsqcup_{Y \subset 3^3} \Omega_Y = \bigsqcup_{w \in S_6^{(3)}} \Omega_w.$$

4.2 Consider a matrix $A \in \Omega_w$ with $w = s_6 s_7 s_3 s_4 s_5 s_6 s_1 s_2 s_3 s_4 s_5$ for $\Omega_w \subset \mathrm{Gr}(3, 8)$.

(a) Find the corresponding Young diagram.
(b) Construct a pipedream and find the elements of the matroid $\mathcal{M}(A)$ appearing in the Young diagram.
(c) Find $\mathcal{M}(A)$ and confirm that the set of indices found in the pipedream, i.e.

$$\{136, 137, 138, 146, 156, 167, 178, 236, 346, 456, 567, 678\},$$

forms a basis $\mathcal{B}(A)$ of the matroid $\mathcal{M}(A)$.

Chapter 5
The Deodhar Decomposition
for the Grassmannian and the Positivity

Abstract In this chapter, we start to review the flag variety G/B^+ for $G = \mathrm{SL}_M(\mathbb{R})$ and the Borel subgroup B^+ of upper triangular matrices, and introduce the Deodhar decomposition of G/B^+ [33, 34]. Then we give a refinement of the Schubert decomposition of $\mathrm{Gr}(N, M)$ as a projection of the Deodhar decomposition, and parametrize each component of the refinement by introducing *Go-diagram*, which is a Young diagram decorated with *black* and *white* stones. In particular, if the Go-diagram has only white stones, it represents a projected Deodhar component of the totally nonnegative (TNN) Grassmannian $\mathrm{Gr}(N, M)_{\geq 0}$. The Go-diagram in this case is the *$\mathcal{J}$-diagram* introduced by Postnikov in [103]. We also construct an explicit form of matrix $A \in \mathrm{Gr}(N, M)_{\geq 0}$ as a point of the projected Deodhar component by using the parameterizations of the flag variety due to Marsh and Rietsch [83]. We conclude this section to give an algorithm to compute an explicit form of the matrix A and to discuss the positivity of A. Most of the materials presented here can be also found in [72].

5.1 The Flag Variety

The following definitions can be made for any split, connected, simply connected, semisimple algebraic group G. However we will be concerned with only the case of $G = \mathrm{SL}_M(\mathbb{R})$.

We fix a maximal torus T, and opposite Borel subgroups B^+ and B^-, which consist of the diagonal, upper-triangular, and lower-triangular matrices, respectively. We let U^+ and U^- be the unipotent radicals of B^+ and B^-, which are the subgroups of upper-triangular and lower-triangular matrices with 1's on the diagonals. For each $1 \leq i \leq M - 1$ we have a homomorphism $\phi_i : \mathrm{SL}_2(\mathbb{R}) \to \mathrm{SL}_M(\mathbb{R})$ such that

© The Author(s) 2017

Y. Kodama, *KP Solitons and the Grassmannians*,

SpringerBriefs in Mathematical Physics 22, DOI 10.1007/978-981-10-4094-8_5

$$\phi_i \begin{pmatrix} a & b \\ c & d \end{pmatrix} = \begin{pmatrix} 1 & & & & \\ & \ddots & & & \\ & & a\ b & & \\ & & c\ d & & \\ & & & \ddots & \\ & & & & 1 \end{pmatrix} \in \mathrm{SL}_M(\mathbb{R}),$$

where the rest of the non-diagonal elements are all zeroes. That is, ϕ_i replaces a 2×2 block of the identity matrix with $\begin{pmatrix} a & b \\ c & d \end{pmatrix}$. Here a is at the $(i+1)$th diagonal entry counting from the southeast corner. (Note that our numbering differs from that in [83] in that the rows of our matrices in $\mathrm{SL}_M(\mathbb{R})$ are numbered from the bottom.)

We use this homomorphism ϕ_i to construct 1-parameter subgroups in G (landing in U^+ and U^-, respectively) defined by

$$x_i(p) = \phi_i \begin{pmatrix} 1 & p \\ 0 & 1 \end{pmatrix} \quad \text{and} \quad y_i(p) = \phi_i \begin{pmatrix} 1 & 0 \\ p & 1 \end{pmatrix} \quad \text{for } p \in \mathbb{R}. \tag{5.1}$$

The datum $(T, B^+, B^-, x_i, y_i; i \in I)$ for G is called a *pinning*.

Let W denote the Weyl group $N_G(T)/T$, where $N_G(T)$ is the normalizer of T. In the case of $G = \mathrm{SL}_M(\mathbb{R})$, we have $W = S_M$ and $T = \{\mathrm{diag}(t_1, \dots, t_M) : t_i \neq 0\}$. The simple reflections $s_i \in S_M$ are given explicitly by $s_i := \dot{s}_i T$ where $\dot{s}_i := \phi_i \begin{pmatrix} 0 & -1 \\ 1 & 0 \end{pmatrix}$ and any $w \in S_M$ can be expressed as a product $w = s_{i_1} s_{i_2} \dots s_{i_m}$ with $m = \ell(w)$ factors. We set $\dot{w} = \dot{s}_{i_1} \dot{s}_{i_2} \dots \dot{s}_{i_m}$.

We have two opposite Bruhat decompositions of the flag variety G/B^+,

$$G/B^+ = \bigsqcup_{w \in W} B^+ \dot{w}\, B^+/B^+ = \bigsqcup_{v \in W} B^- \dot{v}\, B^+/B^+. \tag{5.2}$$

Note that $B^- \dot{v}\, B^+/B^+ \cong \mathbb{R}^{\ell(w_0) - \ell(v)}$ and $B^+ \dot{w}\, B^+/B^+ \cong \mathbb{R}^{\ell(w)}$ where w_0 is the longest element of $W = S_M$. The closure relations for these opposite Bruhat cells are given by $B^- \dot{v}'\, B^+/B^+ \subset \overline{B^- \dot{v}\, B^+/B^+}$ if and only if $v \leq v'$, and also $B^- \dot{w}'\, B^+/B^+ \subset \overline{B^- \dot{w}\, B^+/B^+}$ if and only if $w' \leq w$. We define

$$\mathscr{R}_{v,w} := \left(B^+ \dot{w}\, B^+/B^+ \right) \cap \left(B^- \dot{v}\, B^+/B^+ \right), \tag{5.3}$$

the intersection of opposite Bruhat cells. This intersection is empty unless $v \leq w$, in which case it is smooth of dimension $\ell(w) - \ell(v)$ (see [66, 81]). The strata $\mathscr{R}_{v,w}$ are often called *Richardson varieties*, and the set of those strata $\mathscr{R}_{v,w}$ for $v \leq w$ gives a refinement of the Bruhat decomposition (5.2).

Remark 5.1 One should note that there is another opposite Bruhat intersection defined by $G^{v,w} = B^+ \dot{w} B^+ \cap B^- \dot{v} B^-$, which was studied in [47] for the total positivity of the group G. The strata $\mathscr{R}_{v,w}$ defined in (5.3) should not be mixed with those intersections.

5.1.1   Deodhar Components in the Flag Variety

We now describe the Deodhar decomposition of the flag variety. This is a further refinement of the Bruhat decomposition of G/B^+ into Richardson varieties $\mathscr{R}_{v,w}$. Marsh and Rietsch [83] gave explicit parameterizations for each Deodhar component, identifying each one with a subset in the group.

Definition 5.1 Let $w = s_{i_1} \ldots s_{i_m}$ be a reduced expression for $w \in S_M$ of the length $\ell(w) = m$. We define a symbol $\mathbf{w}$ to emphasize the order of s_i's in the expression and $\mathbf{w} = s_{i_1} \cdots s_{i_m}$ for the fixed order. A *subexpression* $\mathbf{v}$ of $\mathbf{w}$ is defined from the reduced expression $\mathbf{w}$ by replacing some of the factors with 1. That is, $\mathbf{v} = v_1 \cdots v_m$ is a subexpression of $\mathbf{w}$, if $v_l \in \{1, s_{i_l}\}$ for all $l = 1, \ldots, m$. For example, consider a reduced expression $\mathbf{w} = s_3 s_2 s_1 s_3 s_2 s_3 \in S_4$. Then $\mathbf{v} = s_3 s_2 \, 1 \, s_3 s_2 \, 1$ is a subexpression of $\mathbf{w}$. Given a subexpression $\mathbf{v}$, we set $v_{(k)}$ to be the product of the leftmost k factors of $\mathbf{v}$, i.e. $v_{(k)} = v_1 \cdots v_k$, and set $v_{(0)} = 1$ and $v_{(m)} = v$. Thus we use the bold face letter $\mathbf{v}$ to specify the choices $v_l \in \{1, s_{i_l}\}$ in $\mathbf{w}$.

The following definition is given in [83] and is implicit in [33].

Definition 5.2 Given a subexpression $\mathbf{v}$ of a reduced expression $\mathbf{w} = s_{i_1} s_{i_2} \ldots s_{i_m}$, we define

$$
\begin{aligned}
J_{\mathbf{v}}^{\circ} &:= \{k \in \{1, \ldots, m\} \mid v_{(k-1)} < v_{(k)}\}, \\
J_{\mathbf{v}}^{\square} &:= \{k \in \{1, \ldots, m\} \mid v_{(k-1)} = v_{(k)}\}, \\
J_{\mathbf{v}}^{\bullet} &:= \{k \in \{1, \ldots, m\} \mid v_{(k-1)} > v_{(k)}\}.
\end{aligned}
$$

The expression $\mathbf{v}$ is called *non-decreasing* if $v_{(j-1)} \leq v_{(j)}$ for all $j = 1, \ldots, m$, then $J_{\mathbf{v}}^{\bullet} = \emptyset$.

The following definition is from [33, Definition 2.3]:

Definition 5.3 (*Distinguished subexpressions*) A subexpression $\mathbf{v}$ of $\mathbf{w}$ is called *distinguished* if we have

$$
v_{(j)} \leq v_{(j-1)} s_{i_j} \qquad \text{for all } \; j \in \{1, \ldots, m\}. \tag{5.4}
$$

In other words, if right multiplication by s_{i_j} decreases the length of $v_{(j-1)}$, then in a distinguished subexpression we must have $v_{(j)} = v_{(j-1)} s_{i_j}$. We write $\mathbf{v} \prec \mathbf{w}$, if $\mathbf{v}$ is a distinguished subexpression of $\mathbf{w}$.

Example 5.1 Consider $\mathbf{w} = s_2 s_1 s_3 s_2 \in S_4$. Then

(a) The subexpression $\mathbf{v} = s_2 111$ is *not* distinguished, since $v_{(4)} > v_{(3)} s_2$.
(b) The subexpression $\mathbf{v} = 111 s_2$ is distinguished, since $v_{(1)} = v_{(2)} = v_{(3)} < v_{(4)}$.
(c) The subexpression $\mathbf{v} = 1111$ is distinguished, since $v_{(1)} = v_{(2)} = v_{(3)} = v_{(4)}$.
(d) The subexpression $\mathbf{v} = s_2 11 s_2$ is distinguished, since $v_{(1)} = v_{(2)} = v_{(3)} > v_{(4)}$.

Note here that $\mathbf{v}$ in both (c) and (d) gives $v = 1$. Notice that $\mathbf{v}$ in (c) is non-decreasing, but $\mathbf{v}$ in (d) is decreasing. As we will show in Lemma 5.1 that for each $\mathbf{w}$, there is a unique non-decreasing $\mathbf{v} \prec \mathbf{w}$ for the given subexpression $v \leq w$.

Definition 5.4 [83, Definition 5.1] Let $\mathbf{w} = s_{i_1} \ldots s_{i_m}$ be a reduced expression for w, and let $\mathbf{v}$ be a distinguished subexpression. Define a subset $G_{\mathbf{v},\mathbf{w}}$ in G by

$$
G_{\mathbf{v},\mathbf{w}} := \left\{ g = g_1 g_2 \cdots g_m \; \middle| \; \begin{array}{ll} g_\ell = x_{i_\ell}(m_\ell) \dot{s}_{i_\ell}^{-1} & \text{if } \ell \in J_{\mathbf{v}}^{\bullet}, \\ g_\ell = y_{i_\ell}(p_\ell) & \text{if } \ell \in J_{\mathbf{v}}^{\square}, \quad \text{for } p_\ell \in \mathbb{R}^*, \, m_\ell \in \mathbb{R}. \\ g_\ell = \dot{s}_{i_\ell} & \text{if } \ell \in J_{\mathbf{v}}^{\circ}, \end{array} \right\},
$$

(5.5)

where $\mathbb{R}^* = \mathbb{R} \setminus \{0\}$. There is an obvious map $(\mathbb{R}^*)^{|J_{\mathbf{v}}^{\square}|} \times \mathbb{R}^{|J_{\mathbf{v}}^{\bullet}|} \to G_{\mathbf{v},\mathbf{w}}$ defined by the parameters p_ℓ and m_ℓ in (5.5). For $v = w = 1$ we define $G_{\mathbf{v},\mathbf{w}} = \{1\}$.

Example 5.2 Let $W = S_5$, $\mathbf{w} = s_2 s_3 s_4 s_1 s_2 s_3$ and $\mathbf{v} = s_2 111 s_2 1$. Then the corresponding element $g \in G_{\mathbf{v},\mathbf{w}}$ is given by $g = s_2 y_3(p_2) y_4(p_3) y_1(p_4) x_2(m_5) \dot{s}_2^{-1} y_3(p_6)$, which is

$$
g = \begin{pmatrix} 1 & 0 & 0 & 0 & 0 \\ p_3 & 1 & 0 & 0 & 0 \\ 0 & p_6 & 1 & 0 & 0 \\ p_2 p_3 & p_2 - m_5 p_6 & -m_5 & 1 & 0 \\ 0 & -p_4 p_6 & -p_4 & 0 & 1 \end{pmatrix}.
$$

The following result from [83] gives an explicit parametrization for the Deodhar component $\mathscr{R}_{\mathbf{v},\mathbf{w}}$. We will take the description below as the *definition* of $\mathscr{R}_{\mathbf{v},\mathbf{w}}$.

Proposition 5.1 ([83, Proposition 5.2]) *The map* $(\mathbb{R}^*)^{|J_{\mathbf{v}}^{\square}|} \times \mathbb{R}^{|J_{\mathbf{v}}^{\bullet}|} \to G_{\mathbf{v},\mathbf{w}}$ *from Definition 5.4 is an isomorphism. The set* $G_{\mathbf{v},\mathbf{w}}$ *lies in* $U^- \dot{v} \cap B^+ \dot{w} B^+$ *and the assignment* $g \mapsto g \cdot B^+$ *defines an isomorphism*

$$
G_{\mathbf{v},\mathbf{w}} \xrightarrow{\sim} \mathscr{R}_{\mathbf{v},\mathbf{w}} \tag{5.6}
$$

between the subset $G_{\mathbf{v},\mathbf{w}}$ *of the group and the Deodhar component* $\mathscr{R}_{\mathbf{v},\mathbf{w}}$ *in* G/B^+.

For each $w \in S_M$, let us choose a reduced expression $\mathbf{w}$ for w. Then it follows from Deodhar's work (see [33] and [83, Sect. 4.4]) that

$$
\mathscr{R}_{v,w} = \bigsqcup_{\mathbf{v} \prec \mathbf{w}} \mathscr{R}_{\mathbf{v},\mathbf{w}} \quad \text{and} \quad G/B^+ = \bigsqcup_{w \in W} \left(\bigsqcup_{v \leq w} \mathscr{R}_{v,w} \right), \tag{5.7}
$$

where the union $\sqcup_{\mathbf{v} \prec \mathbf{w}}$ is over all distinguished subexpressions for v in $\mathbf{w}$. These are called the *Deodhar decompositions* of $\mathscr{R}_{v,w}$ and G/B^+.

5.1.2 The Totally Nonnegative Part of the Flag Variety

To define the totally nonnegative (TNN) part of the flag variety, we begin with the following definition:

Definition 5.5 [80] The *TNN part* $U_{\geq 0}^-$ of U^- is defined to be the semigroup in U^- generated by the $y_i(p)$ for $p > 0$. The *TNN part* $(G/B^+)_{\geq 0}$ of G/B^+ is defined by

$$(G/B^+)_{\geq 0} := \overline{\{uB^+ \mid u \in U_{\geq 0}^-\}},$$

where the closure is taken inside G/B^+ in its real topology. We sometimes refer to $(G/B^+)_{\geq 0}$ as the *TNN flag variety*.

In [80, 81], Lusztig introduced a natural decomposition of $(G/B^+)_{\geq 0}$.

Definition 5.6 [81] For $v, w \in W$ with $v \leq w$, let

$$\mathscr{R}_{v,w}^{>0} := \mathscr{R}_{v,w} \cap (G/B^+)_{\geq 0}.$$

Then the TNN part of the flag variety G/B^+ has the decomposition

$$(G/B^+)_{\geq 0} = \bigsqcup_{w \in S_M} \left(\bigsqcup_{v \leq w} \mathscr{R}_{v,w}^{>0} \right). \tag{5.8}$$

Lusztig conjectured and Rietsch proved [105] that $\mathscr{R}_{v,w}^{>0}$ is a semi-algebraic cell of dimension $\ell(w) - \ell(v)$. Subsequently Marsh-Rietsch [83] provided an explicit parameterization of each cell. To state their result, we first give the following definition and lemma:

Definition 5.7 (*Positive distinguished subexpressions*) We call a subexpression $\mathbf{v}$ of $\mathbf{w}$ a *positive distinguished subexpression* (or a PDS for short) if

$$v_{(j-1)} < v_{(j-1)}s_{i_j} \quad \text{for all } j \in \{1, \ldots, m\}. \tag{5.9}$$

In other words, it is distinguished and non-decreasing. We write $\mathbf{v}^+$ if $\mathbf{v} \prec \mathbf{w}$ is a PDS.

In Example 5.1, $\mathbf{v} = 1111$ is positive, but $\mathbf{v} = s_2 1 1 s_2$ is not. This example shows that for a subexpression $v = 1 < w$, there is a unique positive subexpression $\mathbf{v} = 1111 \prec \mathbf{w}$. In general, we have the following:

Lemma 5.1 ([83, Lemma 3.5]) *Given $v \le w$ and a reduced expression $\mathbf{w}$ for w, there is a unique $\mathbf{v}^+$ for v in $\mathbf{w}$.*

Proof Let $\mathbf{w} = s_{i_1} \cdots s_{i_m}$ and $\mathbf{v} = v_1 \cdots v_m \prec \mathbf{w}$. The inequality $v_{(l-1)} < v_{(l-1)} s_{i_l}$ implies that $v_{(l-1)}$ cannot have a word ending in s_{i_l}. If $v_{(l)}$ has such an expression, then we must set $v_{(l-1)} = v_{(l)} s_{i_l}$. If it does not, then $v_{(l-1)} = v_{(l)}$. This process defines uniquely the desired positive subexpression of $\mathbf{w}$. □

Their main result is stated as follows:

Theorem 5.1 ([83, Proposition 5.2, Theorem 11.3]) *Choose a reduced expression $\mathbf{w} = s_{i_1} \ldots s_{i_m}$ for w of $\ell(w) = m$. For $v \le w$, we associate the unique PDS $\mathbf{v}^+$ for v in $\mathbf{w}$. Then $J_{\mathbf{v}^+}^{\bullet} = \emptyset$. We define*

$$G_{\mathbf{v}^+, \mathbf{w}}^{>0} := \left\{ g = g_1 g_2 \cdots g_m \left| \begin{array}{ll} g_l = y_{i_l}(p_l) & \text{if } l \in J_{\mathbf{v}^+}^{\square}, \\ g_l = \dot{s}_{i_l} & \text{if } l \in J_{\mathbf{v}^+}^{\circ}, \end{array} \right. \right\}, \tag{5.10}$$

where each p_l is a positive constant denoted by $p_l \in \mathbb{R}_{>0}$. The set $G_{\mathbf{v}^+, \mathbf{w}}^{>0}$ lies in $U^- \dot{v} \cap B^+ \dot{w} B^+$, $G_{\mathbf{v}^+, \mathbf{w}}^{>0} \cong \mathbb{R}_{>0}^{\ell(w)-\ell(v)}$ and the map $g \mapsto gB^+$ defines an isomorphism

$$G_{\mathbf{v}^+, \mathbf{w}}^{>0} \xrightarrow{\sim} \mathscr{R}_{v,w}^{>0}.$$

Example 5.3 Consider the reduced expression $w = s_2 s_3 s_1 s_4 s_5 s_3 s_2 \in S_6$. Let $v = s_3 s_4 s_2 \le w$. Then the PDS $\mathbf{v}^+$ for v is $\mathbf{v}^+ = 1 s_3 1 s_4 1 1 s_2$. That is, $J_{\mathbf{v}^+}^{\square} = \{1, 3, 5, 6\}$ and $J_{\mathbf{v}^+}^{\circ} = \{2, 4, 7\}$. Then the set $G_{\mathbf{v}^+, \mathbf{w}}^{>0}$ consists of all elements of the form

$$y_2(p_1)\dot{s}_3 y_1(p_3)\dot{s}_4 y_5(p_5)y_3(p_6)\dot{s}_2 = \begin{pmatrix} 1 & 0 & 0 & 0 & 0 & 0 \\ p_3 & 0 & -1 & 0 & 0 & 0 \\ p_1 p_3 & 0 & -p_1 & 0 & 1 & 0 \\ 0 & 1 & 0 & 0 & 0 & 0 \\ 0 & p_6 & 0 & 1 & 0 & 0 \\ 0 & 0 & 0 & 0 & p_5 & 1 \end{pmatrix},$$

where each constant $p_i > 0$ for $i \in J_{\mathbf{v}^+}^{\square}$.

5.2 The Deodhar Decomposition for the Grassmannian

The Deodhar decomposition for the Grassmannian $Gr(N, M)$ can be obtained by the projection $\pi_k : G/B^+ \to Gr(N, M)$. Recall that $S_M^{(N)}$ is the set of minimal-length coset representatives of S_N/P_k where P_k is the parabolic subgroup $\langle s_1, \ldots, \widehat{s_{M-N}}, \ldots, s_{M-1} \rangle$ (see Definition 4.4).

For each $w \in S_M^{(N)}$ and $v \le w$, define $\mathscr{P}_{v,w} = \pi_k(\mathscr{R}_{v,w})$. Then by Lusztig [81], π_k is an isomorphism on $\mathscr{P}_{v,w}$ and we have a decomposition

$$\text{Gr}(N, M) = \bigsqcup_{w \in S_M^{(N)}} \left(\bigsqcup_{v \leq w} \mathscr{P}_{v,w} \right). \tag{5.11}$$

That is, each Schubert cell Ω_w for $w \in S_M^{(N)}$ is further decomposed into the components $\mathscr{P}_{v,w}$ with $v \leq w$.

Definition 5.8 For each reduced expression $\mathbf{w}$ for $w \in S_M^{(N)}$, and each $\mathbf{v} \prec \mathbf{w}$, we define the Deodhar component for the Grassmannian, $\mathscr{P}_{\mathbf{v},\mathbf{w}} = \pi_k(\mathscr{R}_{\mathbf{v},\mathbf{w}}) \subset \text{Gr}(N, M)$. Then we have

$$\mathscr{P}_{v,w} = \bigsqcup_{\mathbf{v}} \mathscr{P}_{\mathbf{v},\mathbf{w}}, \tag{5.12}$$

where the union is over all distinguished subexpressions for v in $\mathbf{w}$.

Note that the Deodhar decomposition for the flag variety depends on the choices of reduced expressions $\mathbf{w}$ of each $w \in S_M$. However, its projection to the Grassmannian has a nicer behavior.

Proposition 5.2 *Let $w \in S_M^{(N)}$ and choose a reduced expression $\mathbf{w}$ for w. Then the component $\bigsqcup_{\mathbf{v}} \mathscr{R}_{\mathbf{v},\mathbf{w}}$ does not depend on $\mathbf{w}$, only on w.*

Proof Recall from Lemma 4.1 that any two reduced expressions of $w \in S_M^{(N)}$ can be obtained from each other by a sequence of commuting moves ($s_i s_j = s_j s_i$ where $|i - j| \geq 2$). And it is easy to check that if $s_i s_j = s_j s_i$, then

1. $y_i(a)y_j(b) = y_j(b)y_i(a)$
2. $y_i(a)\dot{s}_j = \dot{s}_j y_i(a)$
3. $(x_i(a)\dot{s}_i^{-1})\dot{s}_j = \dot{s}_j(x_i(a)\dot{s}_i^{-1})$
4. $(x_i(a)\dot{s}_i^{-1})y_j(b) = y_j(b)(x_i(a)\dot{s}_i^{-1})$.

The result now follows from Definition 5.4 and Proposition 5.1. □

Proposition 5.1 gives us a concrete way to construct the projected Deodhar components $\mathscr{P}_{\mathbf{v},\mathbf{w}}$. The projection $\pi_k : G/B \to \text{Gr}(N, M)$ maps each $g \in G_{\mathbf{v},\mathbf{w}}$ to the span of its leftmost k columns:

$$g = \begin{pmatrix} g_{M,M} & \cdots & g_{M,M-N+1} & \cdots & g_{M,1} \\ \vdots & & \vdots & & \vdots \\ g_{1,M} & \cdots & g_{1,M-N+1} & \cdots & g_{1,1} \end{pmatrix} \mapsto A = \begin{pmatrix} g_{1,M-N+1} & \cdots & g_{M,M-N+1} \\ \vdots & & \vdots \\ g_{1,M} & \cdots & g_{M,M} \end{pmatrix}.$$

Alternatively, we may identify $A \in \text{Gr}(N, M)$ with its image in the Plücker embedding. Let e_i denote the column vector in $\mathbb{R}^n$ such that the ith entry from the bottom contains a 1, and all other entries are 0, e.g. $e_M = (1, 0, \ldots, 0)^T$, the transpose of the row vector $(1, 0, \ldots, 0)$. Then the projection π_k maps each $g \in G_{\mathbf{v},\mathbf{w}}$ (identified with $gB^+ \in \mathscr{R}_{\mathbf{v},\mathbf{w}}$) to

$$g \cdot e_{M-N+1} \wedge \ldots \wedge e_M = \sum_{1 \le j_1 < \ldots < j_N \le M} \Delta_{j_1,\ldots,j_N}(A) e_{j_1} \wedge \cdots \wedge e_{j_N}. \qquad (5.13)$$

That is, the Plücker coordinate $\Delta_{j_1,\ldots,j_N}(A)$ is given by

$$\Delta_{j_1,\ldots,j_N}(A) = \langle e_{j_1} \wedge \cdots \wedge e_{j_N}, g \cdot e_{M-N+1} \wedge \cdots \wedge e_M \rangle,$$

where $\langle \cdot, \cdot \rangle$ is the usual inner product on $\wedge^N \mathbb{R}^M$.

One should note that the matrix A is expressed by the set of vectors f_k, where

$$A = (f_1, f_2, \ldots, f_N)^T \quad \text{with} \quad f_k = (g_{1,M-N+k}, \ldots, g_{M,M-N+k})^T.$$

Then one can see the following:

Lemma 5.2 *The matrix A defined by the projection of the matrix g is in a row echelon form.*

Proof Since the pivots are determined by $w \in S_M^{(N)}$, we only need to show this in the case with $v = 11 \cdots 1$. Since g is given by the products of lower triangular matrices $y_i(p)$'s, the vector f_i has the form

$$f_k = (0, \overset{1}{\ldots}, 0, \overset{i_k}{*}, \ldots, *, \overset{M-N+k}{1}, 0, \ldots, 0)^T,$$

where $i_k = w(M - N + k)$ and $*$'s are nonzero elements. Since $g \in U^-$, i.e. the lower triangular matrix with 1's in the diagonals, the pivot set $I = \{i_1, \ldots, i_N\}$ is given by $I = w(M - N + 1, \ldots, M)$. This proves the lemma. □

Example 5.4 We continue Example 5.2 with $\mathbf{w} = s_2 s_3 s_4 s_1 s_2 s_3$ and $\mathbf{v} = s_2 111 s_2 1$. Note that $w \in S_5^{(2)}$. Then the map $\pi_2 : G_{\mathbf{v},\mathbf{w}} \to \mathrm{Gr}(2, 5)$ is given by

$$g = \begin{pmatrix} 1 & 0 & 0 & 0 & 0 \\ p_3 & 1 & 0 & 0 & 0 \\ 0 & p_6 & 1 & 0 & 0 \\ p_2 p_3 & p_2 - m_5 p_6 & -m_5 & 1 & 0 \\ 0 & -p_4 p_6 & -p_4 & 0 & 1 \end{pmatrix} \longrightarrow A = \begin{pmatrix} -p_4 p_6 & p_2 - m_5 p_6 & p_6 & 1 & 0 \\ 0 & p_2 p_3 & 0 & p_3 & 1 \end{pmatrix}.$$

5.2.1 The Go-Diagram

In (5.11), we showed that each Schubert cell Ω_w for $w \in S_M^{(N)}$ in (4.3) has the Deodhar decomposition

$$\Omega_w = \bigsqcup_{v \le w} \mathscr{P}_{v,w}.$$

Recall that the Schubert cell Ω_w can be parametrized by a Young diagram Y_w. Now we parametrize each Deodhar component $\mathscr{P}_{v,w}$ by a *decorated* Young diagram called *Go-diagram* defined as follows:

Definition 5.9 Let $\mathbf{v}$ be a distinguished subexpression of $\mathbf{w} \in S_M^{(N)}$. For each $k \in J_{\mathbf{v}}^\circ$ we place a $\bigcirc$ in the corresponding box of the Young diagram. For each $k \in J_{\mathbf{v}}^\bullet$ we place a $\bullet$ in the corresponding box; and for each $k \in J_{\mathbf{v}}^\square$ we will leave the corresponding box blank. We call the resulting diagram a *Go-diagram* and refer to the symbols $\bigcirc$ and $\bullet$ as *white* and *black stones*.

Example 5.5 For Gr(3, 7), consider the Deodhar component with $(\mathbf{v}, \mathbf{w})$ given by

$$\mathbf{w} = s_3 s_4 s_5 s_6 s_2 s_3 s_4 s_5 s_1 s_2 s_3 s_4, \quad \mathbf{v} = s_3 1 s_5 1 s_2 s_3 1 \overline{s_5} 1 \overline{s_2 s_3} s_4.$$

Note that $\mathbf{w}$ expresses the top cell with the 3×4 rectangular Young diagram. In the subexpression $\mathbf{v}$, the $\overline{s_{i_j}}$ denotes the element which reduces the length, i.e. $v_{(j-1)} > v_{(j)} = v_{(j-1)} s_{i_j}$ for $j \in J_{\mathbf{v}}^\bullet$. Here we have

$$J_{\mathbf{v}}^\circ = \{1, 3, 5, 6, 12\}, \quad J_{\mathbf{v}}^\square = \{2, 4, 7, 9\}, \quad J_{\mathbf{v}}^\bullet = \{8, 10, 11\}.$$

The Young diagram, the reading order and the corresponding Go-diagram are shown in the diagrams below.

s_4	s_3	s_2	s_1
s_5	s_4	s_3	s_2
s_6	s_5	s_4	s_3

12	11	10	9
8	7	6	5
4	3	2	1

Definition 5.10 A Go-diagram is *irreducible* if

(i)   the Young diagram is irreducible (see Definition 4.3), and
(ii) every column and row has at least one empty box.

Note that in terms of the pair (v, w) of the component $\mathscr{P}_{v,w}$, the irreducibility means that $\pi = vw^{-1}$ is a derangement.

If we choose a reading order of O_w, then we will also associate to a Go-diagram of shape O_w a *labeled Go-diagram*, as defined below. Equivalently, a labeled Go-diagram is associated to a pair $(\mathbf{v}, \mathbf{w})$.

Definition 5.11 Given a reading order of O_w and a Go-diagram of shape O_w, we obtain a *labeled Go-diagram* by replacing each $\bigcirc$ with a 1, each $\bullet$ with a -1, and putting a p_i in each blank square b, where the subscript i corresponds to the label of b inherited from the linear extension.

Recall Example 5.5 for Gr(3, 7) with $(\mathbf{v}, \mathbf{w})$ given by

$$\mathbf{w} = s_3 s_4 s_5 s_6 s_2 s_3 s_4 s_5 s_1 s_2 s_3 s_4, \quad \mathbf{v} = s_3 1 s_5 1 s_2 s_3 1 \overline{s_5} 1 \overline{s_2 s_3} s_4 .$$

The permutation $\mathbf{w}$ expresses the top cell of the 3×4 rectangular Young diagram. The reading order and the labeled Go-diagram are given by the diagrams below.

s_4	s_3	s_2	s_1
s_5	s_4	s_3	s_2
s_6	s_5	s_4	s_3

12	11	10	9
8	7	6	5
4	3	2	1

1	-1	-1	p_9
-1	p_7	1	1
p_4	1	p_2	1

Remark 5.2 The Go-diagrams for the distinguished subexpressions in Example 5.1 are shown in the diagrams below.

In particular, notice that the last two diagrams correspond to $v = 1111$ and $v = s_2 1 1 s_2$, respectively, i.e. both give $v = 1$. There are several Go-diagrams with the same choice of $v \leq w$ for different distinguished subexpressions $v \prec w$ (see also Problem 5.2).

5.3 Plücker Coordinates for the Go-Diagram and the Positivity

Consider $\mathscr{P}_{v,w} \subset Gr(N, M)$, where w is a reduced expression for $w \in S_M^{(N)}$ and $v \prec w$. In this section we will provide formulas for the Plücker coordinates $\Delta_I(A)$ of the elements $A \in \mathscr{P}_{v,w}$ whose indices $I \in \mathscr{M}(A)$ are determined directly by the Go-diagram. These formulas are given in terms of the parameters used to define $G_{v,w}$. Some of these formulas are related to corresponding formulas for G/B^+ in [83, Sect. 7].

5.3.1 Formulas for Plücker Coordinates

Let us start with the following lemma.

Lemma 5.3 *Choose any element A of $\mathscr{P}_{v,w} \subset Gr(N, M)$. Let*

$$I = w\{M - N + 1, \ldots, n - 1, n\} \quad and \quad I' = v\{M - N + 1, \ldots, n - 1, n\}.$$

Then if $\Delta_J(A) \neq 0$, we have $I \leq J \leq I'$, where $\leq$ is the lexicographical order from Definition 4.2. In particular, the lexicographically minimal and maximal nonzero Plücker coordinates of A are $\Delta_I(A)$ and $\Delta_{I'}(A)$. Note that if we write $I = \{i_1, \ldots, i_k\}$, then $I' = vw^{-1}\{i_1, \ldots, i_k\}$.

Proof Recall that the component $\mathscr{P}_{v,w}$ is given by $\mathscr{P}_{v,w} = \pi_k(\mathscr{R}_{v,w})$, where $\mathscr{R}_{v,w} \subset \mathscr{R}_{v,w} = (B^+ \dot{w} B^+ / B^+) \cap (B^- \dot{v} B^+ / B^+)$. Now it is easy to check (and is well-known) that the lexicographically minimal minor of each element in the Schubert cell given by the projection $\pi_k(B^+ \dot{w} B^+ / B^+)$ is $\Delta_I(A)$ and the lexicographically maximal minor of each element in the opposite Schubert cell $\pi_k(B^-\dot{v}B^+/B^+)$ is $\Delta_{I'}(A)$ where I and I' are as given above. $\qquad\qquad\qquad\qquad\qquad\qquad\qquad\qquad\Box$

The index set I in Lemma 5.3 represents the set of pivot indices of $A \in \mathscr{P}_{\mathbf{v},\mathbf{w}}$ (see Lemma 5.2). One can also show that the index set $I' = v(M-N+1,\ldots,M)$ gives the last nonzero entry in each row of the matrix A, that is, in each row of A, the first nonzero entry, which is a pivot, is given at $i_k = w(M-N+k)$, and the last nonzero entry is at $i'_k = v(M-N+k) = vw^{-1}(i_k)$.

We now provide formulas for the Plücker coordinates, which are directly determined by the Go-diagram. Before stating a theorem for the formulas, we state the following lemma, which can be easily verified.

Lemma 5.4 *For* $1 \le i \le n-1$, *we have*

(a) $\dot{s}_i e_i = -e_{i+1}, \dot{s}_i e_{i+1} = e_i,$ *and* $\dot{s}_i e_j = e_j$ *if* $j \neq i$ *or* $i+1$.

(b) $y_i(a)e_{i+1} = e_{i+1} + ae_i$ *and* $y_i(a)e_j = e_j$ *if* $j \neq i+1$.

(c) $(x_i(a)\dot{s}_i^{-1})e_i = e_{i+1},\ (x_i(a)\dot{s}_i^{-1})e_{i+1} = -(e_i + ae_{i+1}),$ *and* $(x_i(a)\dot{s}_i^{-1})e_j = e_j$
 for $j \neq i$ *or* $i+1$.

We also fix some notations.

Definition 5.12 Let $\mathbf{w} = s_{i_1}\ldots s_{i_m}$ be a reduced expression for $w \in S_M^{(N)}$ and choose $\mathbf{v} \prec \mathbf{w}$. This determines a Go-diagram D in a Young diagram Y. Let b be any box of D. Note that the set of all boxes of D which are weakly southeast of b forms a Young diagram Y_b^{in}; also the complement of Y_b^{in} in Y is a Young diagram which we call Y_b^{out} (see Example 5.6 below). By looking at the restriction of $\mathbf{w}$ to the positions corresponding to boxes of Y_b^{in}, we obtain a reduced expression $\mathbf{w}_b^{in}$ for some permutation w_b^{in} together with a distinguished subexpression $\mathbf{v}_b^{in}$ for some permutation v_b^{in}. Similarly, by using the positions corresponding to boxes of Y_b^{out}, we obtain $\mathbf{w}_b^{out}, w_b^{out}, \mathbf{v}_b^{out},$ and v_b^{out}. When the box b is understood, we will often omit the subscript b.

For any box b, note that it is always possible to choose a linear extension of O_w, which orders all the boxes of Y^{out} after those of Y^{in}. We can then adjust $\mathbf{w}$ accordingly. Having chosen such a linear extension, we can then write $\mathbf{w} = \mathbf{w}^{in}\mathbf{w}^{out}$ and $\mathbf{v} = \mathbf{v}^{in}\mathbf{v}^{out}$. We then use g^{in} and g^{out} to denote the corresponding factors of $g \in G_{\mathbf{v},\mathbf{w}}$. We define $J_{v^{out}}^\square$ to be the subset of $J_{\mathbf{v}}^\square$ coming from the factors of $\mathbf{v}$ contained in $\mathbf{v}^{out}$, and define $J_{v^{out}}^\circ$ and $J_{v^{out}}^\bullet$ similarly.

Example 5.6 Let $\mathbf{w} = s_4 s_5 s_2 s_3 s_4 s_6 s_5 s_1 s_2 s_3 s_4$ be a reduced expression for $w \in S_7^{(3)}$. Let $\mathbf{v} = s_4 s_5 1 1 s_4 1 s_5 s_1 1 1 s_4$ be a distinguished subexpression. So we have $w = (3,5,6,7,1,2,4)$ and $v = (2,1,3,4,6,5,7)$. The diagrams below represent this data by the poset O_w and the corresponding Go-diagram.

s_4	s_3	s_2	s_1
s_5	s_4	s_3	s_2
s_6	s_5	s_4	

Let b be the box of the Young diagram which is in the second row and the second column (from left to right). Then the diagram below shows the boxes of Y^{in} and Y^{out}, a linear extension which puts the boxes of Y^{out} after those of Y^{in}, and the corresponding labeled Go-diagram. Using this linear extension, $\mathbf{w}^{in} = s_4 s_5 s_2 s_3 s_4$, $\mathbf{w}^{out} = s_6 s_5 s_1 s_2 s_3 s_4$, $\mathbf{v}^{in} = s_4 s_5 1 1 s_4$ and $\mathbf{v}^{out} = 1 s_5 s_1 1 1 s_4$.

out	out	out	out
out	in	in	in
out	in	in	

11	10	9	8
7	5	4	3
6	2	1	

-1	p_{10}	p_9	1
-1	1	p_4	p_3
p_6	1	1	

Note that $J_{\mathbf{v}^{out}}^{\bullet} = \{7, 11\}$ and $J_{\mathbf{v}^{out}}^{\square} = \{6, 9, 10\}$. Then $g \in G_{\mathbf{v},\mathbf{w}}$ has the form

$$g = g^{in} g^{out} = (\dot{s}_4 \dot{s}_5 y_2(p_3) y_3(p_4) \dot{s}_4)\,(y_6(p_6) x_5(m_7) \dot{s}_5^{-1} \dot{s}_1 y_2(p_9) y_3(p_{10}) x_4(m_{11}) s_4^{-1}).$$

When we project the resulting 7×7 matrix to its first three columns, we get the matrix

$$A = \begin{pmatrix} -p_9 p_{10} & -p_3 p_{10} & -p_{10} & -m_{11} & 0 & -1 & 0 \\ 0 & -p_3 p_4 & -p_4 & -m_7 & 1 & 0 & 0 \\ 0 & 0 & 0 & p_6 & 0 & 0 & 1 \end{pmatrix}.$$

Note here that the pivot set is $I = (1, 2, 4) = w(5, 6, 7)$ and the last nonzero entries in the rows are $I' = (6, 5, 7) = v(5, 6, 7)$.

We now state a theorem for formulas for the Plücker coordinates determined by the Go-diagram.

Theorem 5.2 *Let* $\mathbf{w} = s_{i_1} \ldots s_{i_m}$ *be a reduced expression for* $w \in S_M^{(N)}$ *and* $v \prec w$, *and let D be the corresponding Go-diagram. Choose any box b of D, and let* $v^{in} = v_b^{in}$, $w^{in} = w_b^{in}$, $v^{out} = v_b^{out}$, *and* $w^{out} = w_b^{out}$. *Let* $A = \pi_k(g)$ *for any* $g \in G_{\mathbf{v},\mathbf{w}}$, *and let* $I_0 = w\{M - N + 1, \ldots, M - 1, M\}$, *the pivot set. Define* $I_b = v^{in}(w^{in})^{-1} I_0 \in \binom{[M]}{N}$. *If b contains a white or black stone, also define* $I_b^{\circ} = I_b^{\bullet} = v^{in} s_b (w^{in})^{-1} I_0 \in \binom{[M]}{N}$. *If we write* $g = g_1 \ldots g_m$ *as in Definition 5.4, then we have the following:*

(1) $\Delta_{I_b}(A) = (-1)^{|J_{\mathbf{v}^{out}}^{\bullet}|} \prod_{i \in J_{\mathbf{v}^{out}}^{\square}} p_i$.

(2) *If b contains a white stone, then* $\Delta_{I_b^{\circ}}(A) = 0$.

(3) *If b contains a black stone, then* $\Delta_{I_b^{\bullet}}(A) = (-1)^{|J_{\mathbf{v}^{out}}^{\bullet}|+1} m_b \prod_{i \in J_{\mathbf{v}^{out}}^{\square}} p_i + \Delta_{I_b^{\bullet}}(A_b)$, *where m_b is the parameter corresponding to b, and A_b is the matrix A with* $m_b = 0$.

Remark 5.3 The Plücker coordinates given by Theorem 5.2 (1) are monomials in the p_i's. In particular, they are nonzero and do not depend on the values of the m-parameters from the $x_i(m)$-factors. Those minors $\Delta_{I_b}(A)$, $\Delta_{I_b^{\circ}}(A)$ and $\Delta_{I_b^{\bullet}}(A)$ correspond to the chamber minors defined in [83, Definition 6.3]. See also Lemmas

7.4 and 7.5 in [83], and note that the dominant weight for the present case is $\lambda = e_{M-N+1} \wedge \cdots \wedge e_M$.

Also note that if $\mathbf{w}^{out} = \mathbf{w}$ and $\mathbf{v}^{in} = \mathbf{v}$ (respectively $\mathbf{w}^{out} = 1$ and $\mathbf{v}^{out} = 1$), then

$$\Delta_{I_0}(A) = (-1)^{|J_{\mathbf{v}}^{\bullet}|} \prod_{i \in J_{\mathbf{v}}^{\square}} p_i, \qquad (\text{respectively } \Delta_{I'}(A) = 1),$$

where $I = I_0$ and $I' = vw^{-1}I_0$, which are lexicographically minimal and maximal nonzero minors of A (see Lemma 5.3).

Before proving Theorem 5.2, we mention an immediate Corollary.

Corollary 5.1 *Use the notation of Theorem 5.2. Let b be a box of the Go-diagram, and let e, s, and se denote the neighboring boxes, which are at the east, south, and southeast of b, respectively. Then we have*

$$\frac{\Delta_{I_e}(A)\Delta_{I_s}(A)}{\Delta_{I_b}(A)\Delta_{I_{se}}(A)} = \begin{cases} 1 & \text{if box } b \text{ contains a white stone,} \\ -1 & \text{if box } b \text{ contains a black stone,} \\ p_b & \text{if box } b \text{ is blank and the labeled Go diagram contains } p_b. \end{cases}$$

Remark 5.4 Each black and white stone corresponds to a *two-term* Plücker relation, i.e. a three-term Plücker relation in which one term vanishes. Each black stone implies that there are two Plücker coordinates with opposite signs. Also note that the formulas in Corollary 5.1 correspond to the *Generalized Chamber Ansatz* in [83, Theorem 7.1].

Example 5.7 We continue Example 5.6. By Theorem 5.2, $I_0 = w\{5, 6, 7\} = \{1, 2, 4\}$ and $I' = vw^{-1}I_0 = v\{5, 6, 7\} = \{5, 6, 7\}$, and the lexicographically minimal and maximal nonzero Plücker coordinates for A are $\Delta_I(A) = p_3 p_4 p_6 p_9 p_{10}$ and $\Delta_{I'}(A) = 1$; this can be verified for the matrix A above.

We now verify Theorem 5.2 for the box b chosen earlier. Then $I_b = v^{in}(w^{in})^{-1}I = \{1, 4, 6\}$. Theorem 5.2 states that $\Delta_{I_b}(A) = 0$, since this box contains a white stone. The analogous computations for the boxes labeled 7, 6, 4, 3, 2, 1, respectively, yield $\Delta_{1,5,7} = -p_9 p_{10}$, $\Delta_{1,2,7} = p_3 p_4 p_9 p_{10}$, $\Delta_{1,4,5} = p_6 p_9 p_{10}$, $\Delta_{1,3,4} = p_4 p_6 p_9 p_{10}$, $\Delta_{1,2,4} = p_3 p_4 p_6 p_9 p_{10}$, and $\Delta_{1,2,4} = p_3 p_4 p_6 p_9 p_{10}$. These can be checked for the matrix A above.

We now turn to the proof of Theorem 5.2.

Proof For simplicity, we assume that when we write A in row-echelon form, its first pivot is $i_1 = 1$ and its last non-pivot is n, i.e. the corresponding Young diagram is irreducible (see Definition 4.3). (The same proof works without this assumption, but the notation required would be more cumbersome.)

Choose the box b, which is located at the northwest corner of the Young diagram obtained by removing the topmost row and the leftmost column; this is the box labeled 5 in the diagram from Example 5.6. We will explain the proof of the theorem for this box b. The same argument works if b lies in the top row or leftmost column;

and such an argument can be iterated to prove Theorem 5.2 for boxes which are (weakly) southeast of b.

Choose a linear extension of O_w which orders all the boxes of Y^{out} after those of Y^{in}, and which orders the boxes of the top row so that they come after those of the leftmost column. The linear extension from Example 5.6 is one such example. Choosing the reduced expression $\mathbf{w}$ correspondingly, we write $\mathbf{w} = \mathbf{w}^{in}\mathbf{w}^{out}$ and $\mathbf{v} = \mathbf{v}^{in}\mathbf{v}^{out}$, then choose $g \in G_{\mathbf{v},\mathbf{w}}$ and write it as $g = g^{in}g^{out}$. Note that from our choice of linear extension, we have

$$\mathbf{w}^{out} = (s_{M-1}s_{M-2}\ldots s_{M-N+1})(s_1s_2\ldots s_{M-N}). \tag{5.14}$$

Recall that $I_b = v^{in}(w^{in})^{-1}I_0$ where $I_0 = \{i_1, \ldots, i_N\}$, with $i_1 = 1$. In our case, $s_b = s_{M-N}$. Also $w^{-1}I_0 = \{M - N + 1, \ldots, M - 1, M\}$, which implies that

$$(w^{in})^{-1}I_0 = w^{out}\{M - N + 1, \ldots, M - 1, M\} = \{1, M - N + 1, M - N + 2, \ldots, M - 1\}. \tag{5.15}$$

Since there is no factor of s_1 or s_{M-1} in v^{in}, and $I_b = v^{in}\{1, M - N + 1, M - N + 2, \ldots, M - 1\}$, we have

$$1 \in I_b \quad \text{and} \quad M \notin I_b. \tag{5.16}$$

Let us write $I_b = \{j_1, \ldots, j_N\}$ with $j_1 = 1$. Our goal is to compute the minor $\Delta_{I_b}(A) = \langle e_{j_1} \wedge \cdots \wedge e_{j_N}, g \cdot e_{M-N+1} \wedge \cdots \wedge e_M \rangle$.

Let $f_k = g \cdot e_{M-N+k}$. Let q_k be the product of all labels in the "out" boxes of the kth row of the labeled Go-diagram. Using Lemma 5.4 and Eq. (5.14), we obtain

$$f_N = g \cdot e_M = g^{in} \cdot (q_N e_{M-1} + c_M^N e_M),$$
$$f_{N-1} = g \cdot e_{M-1} = g^{in} \cdot (q_{N-1}e_{M-2} + c_{M-1}^{N-1}e_{M-1} + c_M^{N-1}e_M),$$

$$\vdots \qquad \vdots$$

$$f_2 = g \cdot e_{M-N+2} = g^{in} \cdot (q_2 e_{M-N+1} + c_{M-N+2}^2 e_{M-N+2} + \cdots + c_M^2 e_M),$$
$$f_1 = g \cdot e_{M-N+1} = g^{in} \cdot (q_1 e_1 + c_2^1 e_2 + \cdots + c_M^1 e_M).$$

Here the c_i^k's are constants depending on g^{out}.

We now claim that only the first term with coefficient q_k in each f_k contributes to the Plücker coordinate $\Delta_{I_b}(A)$. To prove this claim, note that:

1. Since $M \notin I_b$ and $g^{in} \cdot e_M = e_M$, the terms $c_M^k e_M$ do not affect $\Delta_{I_b}(A)$. Therefore, we may as well assume that each $c_M^\ell = 0$. Define $\tilde{f}_N = q_N g^{in} \cdot e_{M-1}$.
2. Now note that the term $c_{M-1}^{N-1}e_{M-1}$ does not affect the wedge product $\tilde{f}_N \wedge f_{N-1}$. In particular, $\tilde{f}_N \wedge f_{N-1} = \tilde{f}_N \wedge \tilde{f}_{N-1}$ where $\tilde{f}_{N-1} = q_{N-1}g^{in} \cdot e_{M-2}$.

3. Applying the same argument for $2 \leq k \leq N - 2$, we can replace each f_k by $\tilde{f}_k = q_k g^{in} \cdot e_{M-N+k}$ without affecting the wedge product.
4. Since $1 \in I_b$ and e_1 does not appear in any f_k except f_1, for the purpose of computing $\Delta_{I_b}(A)$ we may replace f_1 by $\tilde{f}_1 = q_1 e_1$.

Now we have

$$
\begin{aligned}
\Delta_{I_b}(A) &= \langle e_{j_1} \wedge \cdots \wedge e_{j_N}, \ f_1 \wedge \cdots \wedge f_N \rangle \\
&= \langle e_{j_1} \wedge \cdots \wedge e_{j_N}, \ \tilde{f}_1 \wedge \cdots \wedge \tilde{f}_N \rangle \\
&= \Big(\prod_{j=1}^{N} q_j \Big) \langle e_{j_1} \wedge \cdots \wedge e_{j_N}, \ g^{in} \cdot (e_1 \wedge e_{M-N+1} \wedge \cdots \wedge e_{M-1}) \rangle \\
&= \Big(\prod_{j=1}^{N} q_j \Big) \langle e_{j_2} \wedge \cdots \wedge e_{j_N}, \ g^{in} \cdot (e_{M-N+1} \wedge \cdots \wedge e_{M-1}) \rangle, \quad (5.17)
\end{aligned}
$$

where in the last step we have used $j_1 = 1$. Finally we need to compute the wedge product in (5.17).

From the definition of $I_b = \{j_1, \ldots, j_N\}$, we have $\{j_2, \ldots, j_N\} = v^{in}\{M - N + 1, M - N + 2, \ldots, M - 1\}$. It follows that

$$
\langle e_{j_2} \wedge \cdots \wedge e_{j_N}, \ g^{in} \cdot (e_{M-N+1} \wedge \cdots \wedge e_{M-1}) \rangle = 1,
$$

because this is the lexicographically maximal minor for the matrix $A' = \pi_{N-1}(g^{in}) \in \mathrm{Gr}(N - 1, M - 2)$ corresponding to the sub Go-diagram obtained by removing the top row and leftmost column. Therefore $\Delta_{I_b}(A) = \prod_{j=1}^{k} q_j = (-1)^{|J_{vout}^{\bullet}|} \prod_{i \in J_{vout}^{\square}} p_i$, as desired.

Now consider the case that b contains a white or black stone. Then from the definition of $I_b^{\circ} = I_b^{\bullet} = \{j_1, \ldots, j_N\}$, we have $\{j_2, \ldots, j_N\} = v^{in} s_{M-N}\{M - N + 1, M - N + 2, \ldots, M - 1\}$. The wedge product in (5.17) is equal to $\langle v^{in} s_{M-N} \cdot (e_{M-N+1} \wedge \cdots \wedge e_{M-1}), g^{in} \cdot (e_{M-N+1} \wedge \cdots \wedge e_{M-1}) \rangle$.

If b contains a white stone, then the last factor in v^{in} is s_{M-N} and the last factor in g^{in} is $\dot{s}_{M-N}$, so we can write $v^{in} = \tilde{v}^{in} s_{M-N}$ and $g^{in} = \tilde{g}^{in} \dot{s}_{M-N}$, where $\tilde{v}^{in}$ is also a distinguished expression. Then $\tilde{g}^{in} \in G_{\tilde{v}^{in}, w^{in}}$ so $\tilde{g}^{in} = h \tilde{v}^{in}$ where $h \in U^-$. Then we have $\langle v^{in} s_{M-N} \cdot (e_{M-N+1} \wedge \cdots \wedge e_{M-1}), g^{in} \cdot (e_{M-N+1} \wedge \cdots \wedge e_{M-1}) \rangle = \langle \tilde{v}^{in} \cdot (e_{M-N+1} \wedge \cdots \wedge e_{M-1}), h \tilde{v}^{in} \cdot (e_{M-N+1} \wedge \cdots \wedge e_{M-1}) \rangle$. Since b contains a white stone, $\tilde{v}^{in} s_{M-N} > \tilde{v}^{in}$ in the Bruhat order, and hence $\tilde{v}^{in}\{M - N\} < \tilde{v}^{in}\{M - N + 1\}$. Since $h \in U^-$, it follows that this wedge product equals 0.

If b contains a black stone then the last factor in v^{in} is s_{M-N} and the last two factors in g^{in} are $x_{M-N}(m_b)\dot{s}_{M-N}^{-1}$. So we can write $v^{in} = \tilde{v}^{in} s_{M-N}$ and $g^{in} = \tilde{g}^{in} x_{M-N}(m_b)\dot{s}_{M-N}^{-1}$. Then we have

$$g^{in} \cdot (e_{M-N+1} \wedge \cdots \wedge e_{M-1})$$

$$= \tilde{g}^{in} x_{M-N}(m_b) \dot{s}_{M-N}^{-1} \cdot (e_{M-N+1} \wedge \cdots \wedge e_{M-1}) \tag{5.18}$$

$$= -\tilde{g}^{in} \cdot (m_b(e_{M-N+1} \wedge \cdots \wedge e_{M-1}) + (e_{M-N} \wedge e_{M-N+2} \wedge \cdots \wedge e_{M-1})) \tag{5.19}$$

$$= -m_b \tilde{g}^{in} \cdot (e_{M-N+1} \wedge \cdots \wedge e_{M-1}) - \tilde{g}^{in} \cdot (e_{M-N} \wedge e_{M-N+2} \wedge \cdots \wedge e_{M-1}). \tag{5.20}$$

Note that to go from (5.18) to (5.19) we have used Lemma 5.4.

Let us compute the wedge product of the first term in (5.20) with $v^{in} s_{M-N} \cdot (e_{M-N+1} \wedge \cdots \wedge e_{M-1})$. Using $v^{in} = \tilde{v}^{in} s_{M-N}$, this can be expressed as

$$-m_b \cdot \langle v^{in} \cdot (e_{M-N} \wedge e_{M-N+2} \wedge \cdots \wedge e_{M-1}), \tilde{g}^{in} \cdot (e_{M-N+1} \wedge \cdots \wedge e_{M-1}) \rangle$$

$$= -m_b \cdot \langle \tilde{v}^{in} \cdot (e_{M-N+1} \wedge \cdots \wedge e_{M-1}), \tilde{g}^{in} \cdot (e_{M-N+1} \wedge \cdots \wedge e_{M-1}) \rangle.$$

Since we again have $\tilde{g}^{in} = h\tilde{v}^{in}$ where $h \in U^-$, the above quantity equals $-m_b$.

Let us now compute the wedge product of the second term in (5.20) with $v^{in} s_{M-N} \cdot (e_{M-N+1} \wedge \cdots \wedge e_{M-1})$. This wedge product can be written as

$$\langle v^{in} \cdot (e_{M-N} \wedge e_{M-N+2} \wedge \cdots \wedge e_{M-1}), \tilde{g}^{in} \cdot (e_{M-N} \wedge e_{M-N+2} \wedge \cdots \wedge e_{M-1}) \rangle$$

$$= \langle v^{in} \cdot (e_{M-N} \wedge e_{M-N+2} \wedge \cdots \wedge e_{M-1}), \tilde{g}^{in} \dot{s}_{M-N}^{-1} \cdot (e_{M-N+1} \wedge \cdots \wedge e_{M-1}) \rangle$$

$$= \Delta_{j_1,\ldots,j_N}(A_b),$$

where A_b is the matrix obtained from A by setting $m_b = 0$. This completes the proof of the theorem. $\qquad \square$

5.3.2 The Le-Diagrams and the Pipedreams

We can use the previous results on the Plücker coordinates to obtain *nonnegative* Deodhar components in the Grassmannian.

Theorem 5.3 *Consider $A \in \mathrm{Gr}(N, M)$ lying in some Deodhar component $\mathscr{P}_{v,w}$ with the Go-diagram D. Consider the collection $\mathscr{J} = \{\Delta_{I_0}(A)\} \cup \{\Delta_{I_b}(A) \mid b$ a box of $D\}$, where I_0 and I_b are defined as in Theorem 5.2. If all of these minors are positive, then D has no black stones, and all of the parameters p_i must be positive. It follows that A lies in $\mathscr{P}_{v,w}^{>0} \subset \mathrm{Gr}(N, M)_{\geq 0}$.*

Proof By Remark 5.4, if all the minors in $\mathscr{J}$ are positive, then D cannot have a black stone.

By Theorem 5.2, we have that

$$\Delta_{I_0}(A) = (-1)^{|J_v^{\bullet}|} \prod_{i \in J_v^{\square}} p_i \quad \text{and} \quad \Delta_{I_b}(A) = (-1)^{|J_{vout}^{\bullet}|} \prod_{i \in J_{vout}^{\square}} p_i.$$

Since we are assuming that both of these are positive, it follows that for any box b,

$$\frac{\Delta_{I_0}(A)}{\Delta_{I_b}(A)} = (-1)^{|J^\bullet_{\sqrt{in}}|} \prod_{i \in J^\square_{\sqrt{in}}} p_i$$

is also positive. Now by considering the boxes b of D in an order proceeding from southeast to northwest, it is clear that every parameter p_i in the labeled Go-diagram must be positive, because each $\frac{\Delta_{I_0}(A)}{\Delta_{I_b}(A)}$ must be positive.   □

Definition 5.13 A Go-diagram without ● is the $\cal{J}$-*diagram* introduced by Postnikov in [103], i.e. the corresponding subexpression is *positive*. The $\cal{J}$-diagram is characterized by the property (called $\cal{J}$-*property*): If there is ○, then all the boxes either to its left or above it are all ○. That is, there is no such ○ which has an empty box to its left and an empty box above it.

Then we have:

Theorem 5.4 ([103]) *There is a bijection between the set of* irreducible $\cal{J}$-diagrams *(i.e. irreducible Go-diagrams without* ●*) and the set of* derangements *of S_M. That is, each TNN (irreducible) Deodhar component $\mathscr{P}^{>0}_{v,w}$ in* $\mathrm{Gr}(N, M)_{\geq 0}$ *can be parametrized by a unique (irreducible) $\cal{J}$-diagram.*

Thus, we have the Deodhar decomposition for the TNN Grassmannian $\mathrm{Gr}(N, M)_{\geq 0}$,

$$\mathrm{Gr}(N, M)_{\geq 0} = \bigsqcup_{w \in S^{(N)}_M} \Omega^{>0}_w \quad \text{with} \quad \Omega^{>0}_w = \bigsqcup_{v \leq w} \mathscr{P}^{>0}_{v,w},$$

where each v is a positive subexpression of w and each irreducible component $\mathscr{P}^{>0}_{v,w}$ can be uniquely parametrized by a derangement of S_M.

We here remark that each derangement associated to the $\cal{J}$-diagram can be found by constructing a pipedream on the diagram (see Sect. 4.3.2). Starting from a $\cal{J}$-diagram, we replace a blank box with a box containing elbow-pipes connected by a bridge and replace a box with white stone by a box containing crossing pipes as shown below.

Then we label the southeast (input) boundary of the $\cal{J}$-diagram from 1 to M starting from the top corner to the bottom corner of the boundary. We place a pipe with the index of the input edge from the southeast (output) boundary to the northwest boundary, and then label each northwest edge according to the index of the pipe. Then the derangement π is given by $\pi = vw^{-1}$, and a pair of indices (i, j) with $\pi(i) = j$ can be found on the opposite sides of the boundary.

Example 5.8 Consider the case with $\pi = (7, 4, 2, 9, 1, 3, 8, 6, 5)$. The pair (v, w) is given by

$$w = s_7 s_8 s_4 s_5 s_6 s_7 s_2 s_3 s_4 s_5 s_6 s_1 s_2 s_3 s_4 s_5, \qquad v = s_8 s_4 s_5 s_6 s_1 s_4.$$

The J-diagram is then given by the left panel in the figure below.

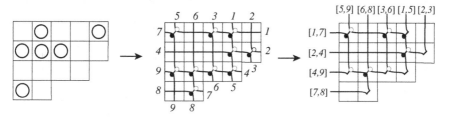

The right diagram shows the *reduced* pipedream, and each pipe has the index pair $[i, j]$ representing a permutation of the indices (i, j).

Using the pipedream, one can also obtain some of the elements in $\mathcal{M}(A)$. For the above example, we can construct the diagrams below. The left diagram shows the reading order of $w \in S_9^{(4)}$. The middle one shows the elements of $\mathcal{M}(A)$ appearing in the diagram, and each element I_b at the box b is given by the $v^{in}(w^{in})^{-1}I_0$ with the pivot set $I_0 := \{1, 2, 4, 7\}$ where the pair (v^{in}, w^{in}) for the box b is defined in Definition 5.12.

16	15	14	13	12
11	10	9	8	7
6	5	4	3	
2	1			

4789	4578	4567	3457	1347
1489	1478	1467	1457	1347
1289	1278	1267	1256	
1248	1248			

p_{16}	1	p_{14}	p_{13}	1
1	1	1	p_8	p_7
p_6	p_5	p_4	p_3	
1	p_1			

The right diagram shows the *labeled* J-diagram with the parameters p_i's. Then the Plücker coordinates can be calculated using the formula in Theorem 5.2. For example, at the box labeled by 10, we have $\Delta_{1478} = p_6 p_{13} p_{14} p_{16}$. Since the matrix A can be constructed by those parameters, we need only a subset of $\mathcal{M}(A)$ determined by the boxes of the diagram with p_i's. That is, in this example, the set $\mathcal{T}(A) \subset \mathcal{M}(A)$ is given by

$$\mathcal{T}(A) = \{I = 1247, 1248, 1256, 1267, 1278, 1289, 1347, 1457, 3457, 4567, 4789\}.$$

This is the result obtained in [124] (see also [73]).

Remark 5.5 Here we make some remarks on the vanishing minors for $A \in \mathscr{P}_{\mathbf{v},\mathbf{w}}^{>0} \subset$ $\mathrm{Gr}(N, M)_{\geq 0}$. From the pipedream, one can compute the index set of the Plücker coordinate for each box labeled by b, which is given by $I_b := v_b^{in}(w_b^{in})^{-1}I_0$ with the pivot set I_0. Let $\{i_1, \ldots, i_l\}$ be the pivots corresponding to the l rows above the box b, and let $\{i_{N-m+1}, \ldots, i_N\}$ be the pivots in the m columns left of the box b. One can easily show that those pivots are the fixed points of the permutation $\sigma_b := v_b^{in}(w_b^{in})^{-1}$, i.e. $\sigma_b(i_k) = i_k$ for $k = 1, \ldots, l$ and $k = i_{N-m+1}, \ldots, i_N$. Define $\{a_1, \ldots, a_n\}$ as the consecutive numbers from $a_1 = i_{l+1}$ to $a_n < i_{N-m+1}$ where a_n is given by the input index of the south boundary of the column containing the box b and i_{N-m+1} is the smallest pivot index larger than a_n. This gives us

$$i_1 = 1 < i_2 < \cdots < i_l < a_1 < \cdots < a_n < i_{N-m+1} < \cdots < i_N < M.$$

For fixed index sets $\{i_1, \ldots, i_l\}$ and $\{i_{N-m+1}, \ldots, i_N\}$, we let $\mathscr{M}_b(A)$ denote the subset of the matroid $\mathscr{M}(A)$ defined by

$$\mathscr{M}_b(A) := \left\{ \{i_1, \ldots, i_l, b_1, \ldots, b_p, i_{N-m+1}, \ldots, i_N\} : b_k \in \{a_1, \ldots, a_n\} \right\},$$

where $k = 1, \ldots, p = N - (m + l)$. Then $I_b = \sigma_b I_0 = v_b^{in}(w_b^{in})^{-1}I_0$ is the lexicographically maximal element of $\mathscr{M}_b(A)$, which is given by

$$I_b = \{i_1, \ldots, i_l, \sigma_b(i_{l+1}), \ldots, \sigma_b(i_{N-m}), i_{N-m+1}, \ldots, i_N\}.$$

Then we have

$$\Delta_J(A) = 0 \quad \text{for all} \quad J = \{i_1, \ldots, i_l, c_1, \ldots, c_p, i_{N-m+1}, \ldots, i_N\} > I_b.$$

This will be useful when we compute the dominant phase $\Theta_I(x, y, t)$ in the τ-function. Notice that the vanishing Plücker coordinates in Theorem 5.2 can be confirmed by this remark.

Before closing this section, we mention the following proposition which is easy to verify (Problem 5.5):

Proposition 5.3 *Let A be an irreducible element in $\mathscr{P}_{\mathbf{v},\mathbf{w}}^{>0} \subset \mathrm{Gr}(N, M)_{\geq 0}$ with dimension d, i.e. $d = \ell(vw^{-1})$. Then the reduced pipedream for A has the following structure:*

(a) *The number of white vertices is given by $d - N$.*
(b) *The number of black vertices is given by $d - (M - N)$.*

As will be shown in Chap. 8, those trivalent vertices will be identified as the resonant interaction points in the soliton graph.

5.3.3 A Network Representation of the Element $g \in G_{v,w}$

Here we give a network representation of the pair $(\mathbf{v}, \mathbf{w})$ to compute the element $g \in G_{v,w}$ and the matrix $A \in \mathrm{Gr}(N, M)$. The network representation was originally invented for a parametrization of totally positive matrices in [13], and it was generalized in [83]. In this section, we consider the network representation in [83] for only the cases with the positive subexpressions (see [127] for the general case).

For the index set (i, j) of the entry $g_{i,j}$ of the matrix g, one can interpret i as the input index and j as the output index in a network representation of g. To be more precise, we give the following networks for the matrices $y_i(p)$ and $\dot{s}_i$ in (5.1) as shown below.

$$\text{for} \quad y_i(p) = \phi_i \begin{pmatrix} 1 & 0 \\ p & 1 \end{pmatrix}$$

$$\text{for} \quad s_i = \phi_i \begin{pmatrix} 0 & -1 \\ 1 & 0 \end{pmatrix}$$

Here the labels at the left side are the input (*row*) indices and the right ones are the output (*column*) indices. The arrow in each edge shows the direction of path from the input to the out put. The indices $j \neq i, i + 1$ are omitted in the figure and are connected with the lines having the same indices at both ends, i.e. j to j indicating the diagonal elements in y_i and s_i except $j = i, i + 1$. The middle connection marked with the parameter p in the $y_i(p)$-network indicates the entry $g_{i,i+1}$ (recall that the entry $g_{i,j}$ is numbered from the bottom-right corner of matrix g), and can only connect from the white vertex to the black vertex as shown by the arrow in the network. Since every arrow is defined as pointing towards the right, we omit the arrows in the following network diagrams.

We then define the network associated with the matrix $g \in G_{v,w}$ as a concatenation of the networks of $y_i(p)$'s and s_i's. Let us explain this with the example of $\mathrm{Gr}(4, 9)$ from the end of the previous section. The pair $(\mathbf{v}, \mathbf{w})$ is given by

$$\mathbf{w} = s_7 s_8 s_4 s_5 s_6 s_7 s_2 s_3 s_4 s_5 s_6 s_1 s_2 s_3 s_4 s_5, \qquad \mathbf{v} = 1 \, s_8 \, 1 \, 1 \, 1 \, 1 \, 1 \, s_4 s_5 s_6 s_1 \, 1 \, 1 \, s_4 \, 1 \, .$$

We sometimes express $(\mathbf{v}, \mathbf{w})$ simply as $s_7 \overline{s_8} s_4 s_5 s_6 s_7 s_2 s_3 \overline{s_4 s_5 s_6 s_1} s_2 s_3 \overline{s_4} s_5$. The matrix g is then given by

$$g = y_7(p_1) \dot{s}_8 y_4(p_3) y_5(p_4) y_6(p_5) y_7(p_6) y_2(p_7) y_3(p_8) \dot{s}_4 \dot{s}_5 \dot{s}_6 \dot{s}_1 y_2(p_{13}) y_3(p_{14}) \dot{s}_4 y_5(p_{16}).$$

The network corresponding to g is then given by

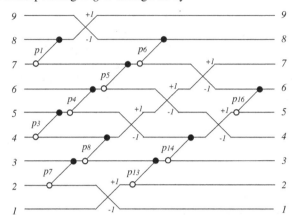

and the matrix element $g_{i,j}$ can be computed as follows: Consider a path P from the input index i to the output index j in the network and define the weight $w(P)$ as the product of all the weights along the path P. Then we have

$$g_{i,j} = \sum_P w(P),$$

where the sum is over all possible paths from i to j. The g-matrix is found to be

$$g = \begin{pmatrix} 0 & -1 & 0 & 0 & 0 & 0 & 0 & 0 & 0 \\ 1 & 0 & 0 & 0 & 0 & 0 & 0 & 0 & 0 \\ p_1 & p_6 & 0 & -1 & 0 & 0 & 0 & 0 & 0 \\ 0 & p_5p_6 & 0 & -p_5 & 0 & 1 & 0 & 0 & 0 \\ 0 & p_4p_5p_6 & 0 & -p_{16}-p_4p_5 & -1 & p_4 & 0 & 0 & 0 \\ 0 & p_3p_4p_5p_6 & 1 & -p_3p_{16}-p_3p_4p_5 & -p_3 & p_3p_4 & 0 & 0 & 0 \\ 0 & 0 & p_8 & p_{14}p_{16} & p_{14} & 0 & 1 & 0 & 0 \\ 0 & 0 & p_7p_8 & p_7p_{14}p_{16} & p_7p_{14} & 0 & p_7 & 0 & -1 \\ 0 & 0 & 0 & p_{13}p_{14}p_{16} & p_{13}p_{14} & 0 & p_{13} & 1 & 0 \end{pmatrix}$$

Recall that the numbering of the entries are counted from the southeast corner. For example, the top-left (or northwest) corner of the matrix is $g_{9,9}$, and it has no path from 9 to 9, i.e. $g_{9,9} = 0$. Then the matrix A is given by the first four columns of g, and we obtain

$$A = \begin{pmatrix} p_{13}p_{14}p_{16} & p_7p_{14}p_{16} & p_{14}p_{16} & -p_3q_{16} & -q_{16} & -p_5 & -1 & 0 & 0 \\ 0 & p_7p_8 & p_8 & 1 & 0 & 0 & 0 & 0 & 0 \\ 0 & 0 & 0 & p_3p_4p_5p_6 & p_4p_5p_6 & p_5p_6 & p_6 & 0 & -1 \\ 0 & 0 & 0 & 0 & 0 & 0 & p_1 & 1 & 0 \end{pmatrix}$$

with $q_{16} = p_{16} + p_4p_5$. Confirm that the pivot set is given by $I = \{1, 2, 4, 7\} = w\{6, 7, 8, 9\}$ and the last nonzero entries are at $I' = \{7, 4, 9, 8\} = v\{6, 7, 8, 9\}$.

Remark 5.6 The network defined above is actually the same as the pipedream on the $\mathcal{J}$-diagram (see also [103]). We show this by using the previous example. Let us

first recall the pipedreams on each box in the $\bot$-diagram, i.e. we put the elbow pipes
and a bridge for empty boxes, and crossing pipes for boxes with white stones (or 0)
as shown below.

Here, the east and south edges give the input indices while the west and north give
the output indices. Now compare the pipedreams in Example 5.8 with the networks
given above (note that the output indices indicate the labels of the pipes marked by
the input indices).

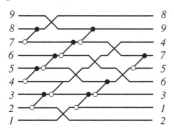 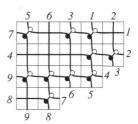

Here the input indices on the west side of the network appear in the southeast boundary of the Young diagram, and the output indices on the east side of the network
appear in the northwest boundary. Also note that the black-white vertices are located
at the same crossing locations in both diagrams.

5.3.4 The Chord Diagrams and Their Combinatorics

Recall that each matrix $A \in \mathrm{Gr}(N, M)_{\geq 0}$ is uniquely parametrized by a $\bot$-diagram
associated with a pair of permutations (v, w), such that $v \leq w$ is a positive subexpression of $w \in S_M^{(N)}$. In particular, the permutation $\pi = vw^{-1}$ is a derangement
in S_M for the irreducible $\bot$-diagram. Here we represent each derangement $\pi =
(\pi(1), \pi(2), \ldots, \pi(M))$ by a chord diagram defined below and give a formula for
the number of irreducible $\bot$-diagrams, i.e. derangements in S_M.

Definition 5.14 A chord diagram associated with a derangement $\pi \in S_M$ is
defined as follows: Consider a line segment with M marked points by the number
$\{1, 2, \ldots, M\}$ in increasing order from left to right.

(a) If $i < \pi(i)$ (excedance), then draw a chord joining i and $\pi(i)$ on the upper part
of the line.
(b) If $j > \pi(j)$ (deficiency), then draw a chord joining j and $\pi(j)$ on the lower part
of the line.

For Example 5.8, we have $\pi = (7, 4, 2, 9, 1, 3, 8, 6, 5)$ and the figure below shows corresponding chord diagram.

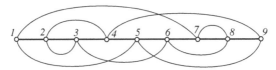

Each $\mathcal{J}$-diagram associated to a derangement $\pi \in S_M$ gives a unique parametrization of an irreducible Deodhar component $\mathscr{P}_{v,w}^{>0}$ in $\mathrm{Gr}(N, M)_{\geq 0}$, whose dimension can be computed from the corresponding chord diagram [103, 133]. It was shown in [31] that the dimension of the component $\mathscr{P}_{v,w}^{>0}$ is given by

$$\dim (\mathscr{P}_{v,w}^{>0}) = N + C(\pi),$$

where $C(\pi)$ is the number of crossings in the chord diagram and the crossings are defined by $C(\pi) = C_+(\pi) + C_-(\pi)$ where

$$C_\pm(\pi) = \sum_{i=1}^{M} C_\pm(i), \quad \text{with} \quad \begin{cases} C_+(i) := |\{j : j < i < \pi(j) < \pi(i)\}| , \\ C_-(i) := |\{j : j > i \geq \pi(j) > \pi(i)\}| . \end{cases} \quad (5.21)$$

(Note that the above definitions of $C_\pm(i)$ are switched from those given in Ref. [31]). For the above example, the chord diagram associated to $\pi = (7, 4, 2, 9, 1, 3, 8, 6, 5)$ has $C_+(\pi) = 1, C_-(\pi) = 5$ and the corresponding component $\mathscr{P}_{v,w}^{>0}$ in $\mathrm{Gr}(4, 9)_{\geq 0}$ has dimension $10 = 4 + 1 + 5$.

Let $\mathscr{D}_M \subset S_M$ be the set of all derangements in S_M. Then the derangements can be enumerated according to the number of excedances $e(\pi)$ of π by the generating polynomial called the derangement polynomial,

$$D_M(p) = \sum_{\pi \in \mathscr{D}_M} p^{e(\pi)} = \sum_{N=1}^{M-1} D_{N,M} p^N, \quad \text{for} \quad M \geq 1,$$

where the coefficient $D_{N,M}$ denotes the number of derangements of $[M]$ with N excedances. The total number of derangements in S_M is then given by $|\mathscr{D}_M| = D_M(1)$. The explicit formula for the derangement polynomial $D_M(p)$ is obtained from the exponential generating function [108] with $D_0(p) := 1$,

$$D(p, z) = \sum_{M=0}^{\infty} D_M(p) \frac{z^M}{M!} = \frac{1 - p}{e^{zp} - p e^z} . \quad (5.22)$$

The first few polynomials are given by $D_1(p) = 0$, $D_2(p) = p$, $D_3(p) = p + p^2$, $D_4(p) = p + 7p^2 + p^3$. Moreover, the polynomials $D_M(p)$ are *symmetric*, i.e. its coefficients satisfy the following relation

$$D_{N,M} = D_{M-N,M}, \quad \text{for} \quad N = 1, 2, \ldots, M - 1. \tag{5.23}$$

This and various other properties of the derangement polynomial $D_M(p)$ can be found in [108]. Note that Eq. (5.23) is equivalent to the relation

$$p^M D_M(p^{-1}) = D_M(p),$$

which in turn follows from the symmetry

$$D(p^{-1}, zp) = D(p, z)$$

of the exponential generating function, and can be verified directly from Eq. (5.22).

The above formulas give the number $D_{N,M}$ of the irreducible component $\mathscr{P}^{>0}_{v,w} \subset \mathrm{Gr}(N, M)_{\geq 0}$. Thus, when $M = 3$, there are only *two* cases corresponding to $\pi = (3, 1, 2)$ for $\mathrm{Gr}(1, 3)_{\geq 0}$ and $\pi = (2, 1, 3)$ for $\mathrm{Gr}(2, 3)_{\geq 0}$. When $M = 4$, there are *seven* cases in $\mathrm{Gr}(2, 4)_{\geq 0}$ (see Problem 5.3). It is also possible to obtain a further refinement of the total number of irreducible component $\mathscr{P}^{>0}_{v,w}$ by introducing a q-analog of the derangement number $D_{N,M}$ namely,

$$D_{N,M}(q) = \sum_{r=N}^{N(M-N)} D_{r,N,M}\, q^r,$$

where $D_{r,N,M}$ is the number of derangements of $[M]$ with N excedances, and $r - N$ crossings as defined in Eq. (5.21). Then, $D_{N,M}(q)$ is the generating polynomial for the irreducible components in $\mathrm{Gr}(N, M)_{\geq 0}$, and $D_{r,N,M}$ is the number of those components of dimension r. The upper limit $N(M - N)$ in the sum gives the maximum dimensional component in $\mathrm{Gr}(N, M)_{\geq 0}$. That is, the total number of irreducible $\mathcal{J}$-diagrams is given by $D_{N,M}(q = 1) = D_{N,M}$.

It is interesting to note that $D_{N,M}(q)$ is related to a q-analog of the Eulerian number [133],

$$E_{k,n}(q) = q^{n-k^2-k} \sum_{i=0}^{k-1} (-1)^i\, [k-i]_q^n\, q^{ki} \left(\binom{n}{i} q^{k-i} + \binom{n}{i-1} \right),$$

where $[k]_q := \frac{1-q^k}{1-q} = 1 + q + q^2 + \cdots + q^{k-1}$ is the q-analog of the number k. The polynomial $E_{k,n}(q)$ was recently introduced in [133], where a rank generating function for the cells in $\mathrm{Gr}(N, M)_{\geq 0}$ was derived by building on the work of [103]. It follows from [31, 133] that the coefficient of q^r in $E_{N,M}(q)$ is the number of permutations of S_M with N *weak* exedances and whose chord diagrams have $r - N$ crossings as defined in Eq. (5.21). Recall that $l \in [M]$ is called a *weak* exedance of a permutation $\pi \in S_M$ if $\pi(l) \geq l$. The following result gives the relation between the polynomials $D_{N,M}(q)$ and $E_{N,M}(q)$.

Proposition 5.4 *The generating polynomial for the number of derangements of $[M]$ is given by*

$$D_{N,M}(q) = \sum_{j=0}^{N-1} (-1)^j \binom{M}{j} E_{N-j,M-j}(q),$$

where $E_{k,n}(q)$ are the Eulerian polynomials defined above.

Proof Let $E(N, M)$ denote the set of all permutations of S_M with N weak exedances, and let $D(N, M) \subset \mathscr{D}_M$ denote the derangements of $[M]$ with N exedances. Then the polynomials $E_{N,M}(q)$ and $D_{N,M}(q)$ are given in terms of the number of crossings $C(\pi)$ of the permutation π as

$$E_{N,M}(q) = q^N \sum_{\pi \in E(N,M)} q^{C(\pi)}, \qquad D_{N,M}(q) = q^N \sum_{\pi \in D(N,M)} q^{C(\pi)}.$$

Note that for each $n \leq N - 1$, an element of $E(N, M)$ can be obtained by adding n fixed points to the corresponding element of the derangement set $D(N - n, M - n)$. Then for $N' = N - n$, $M' = M - n$,

$$E_{N,M}(q) = q^N \sum_{\substack{S \subset [M], \\ |S|=n}} \sum_{\pi \in D(N',M')} q^{C(\pi)} \;=\; \sum_{n=0}^{N-1} \binom{M}{n} D_{N-n,M-n}(q).$$

Inverting the above formula yields the desired result.   □

Proposition 5.4 provides an explicit formula for enumerating the irreducible components associated to the derangements $\pi \in S_M$ with specific dimensions. For example, when $M = 4$ and $N = 2$, Proposition 5.4 yields $D_{2,4}(q) = q^4 + 4q^3 + 2q^2$. This implies that there is one component of dimension 4 (top one), four cells of dimension 3, and two components of dimension 2.

Problems

5.1 Find all the irreducible Go-diagrams for $Gr(2, 4)$.

5.2 Find all the irreducible Go-diagrams for the top cell of $Gr(3, 6)$ with $v = id$. (Need to find all the distinguished subexpressions with $v = id$ for $w = s_3 s_4 s_5 s_2 s_3 s_4 s_1 s_2 s_3$.)

5.3 Show that there are seven cases for irreducible matrix $A \in Gr(2, 4)_{\geq 0}$. Find the $\mathscr{J}$-diagrams for all those cases.

5.4 Let π be a derangement given by

$$\pi = \begin{pmatrix} 1\,2\,3\,4\,5\,6\,7\,8 \\ 3\,1\,7\,6\,2\,8\,5\,4 \end{pmatrix}.$$

(a) Find the Young diagram and $w \in S_8^{(3)}$.
(b) Find the J-diagram.
(c) Find the corresponding pair (v, w) where v is the positive subexpression of w.
(d) Find the Plücker coordinates which are determined by the J-diagram.
(e) Find the A-matrix.

5.5 Prove Proposition 5.3 on the numbers of white and black vertices in the reduced pipedream on any J-diagram.

Chapter 6
Classification of KP Solitons

Abstract This chapter concerns with a *classification* problem of the asymptotic spatial structure of "regular" KP solitons for large $|y|$ using the Deodhar decomposition of the Grassmannian. More precisely, we consider the soliton solution from a point A represented by an $N \times M$ matrix on a Deodhar component $\mathscr{P}_{v,w}^{>0}$ of the TNN Grassmannian $\mathrm{Gr}(N, M)_{\geq 0}$ and classify the asymptotic structure of the solution in the xy-plane. It turns out that the asymptotic structure of the soliton solutions from the matrix A is completely characterized by a derangement $\pi = vw^{-1}$ of the Deodhar component $\mathscr{P}_{v,w}^{>0}$, such that the line-solitons for $y \gg 0$ are of $[i, \pi(i)]$-types with $i < \pi(i)$ and those for $y \ll 0$ are of $[\pi(j), j]$-types with $\pi(j) < j$.

6.1 Soliton Graphs $\mathscr{C}_t(\mathscr{M}(A))$

Recall that the KP soliton is expressed by the form

$$u(x, y, t) = 2\partial_x^2 \ln \tau(x, y, t),$$

where the τ-function is given by

$$\tau(x, y, t) = \sum_{J \in \mathscr{M}(A)} \Delta_J(A) E_J(x, y, t). \tag{6.1}$$

Here, $J = \{j_1, \ldots, j_N\}$ and $E_J = \mathrm{Wr}(E_{j_1}, \ldots, E_{j_N}) = K_J \exp \Theta_J$ with

$$\begin{cases} K_J = \displaystyle\prod_{l>m}^{N} (\kappa_{j_l} - \kappa_{j_m}), \\ \Theta_J = \displaystyle\sum_{n=1}^{N} \theta_{j_n} \quad \text{with} \quad \theta_j := \kappa_j x + \kappa_j^2 y + \kappa_j^3 t. \end{cases}$$

The asymptotic spatial structure of the solution $u(x, y, t)$ is determined from the consideration of which exponential term $E_J(x, y, t)$ in the τ-function dominates in different regions of the xy-plane for large $|y|$. For example, if only one exponential term E_J in the τ-function is dominant in a certain region, then the solution

© The Author(s) 2017

Y. Kodama, *KP Solitons and the Grassmannians*,

SpringerBriefs in Mathematical Physics 22, DOI 10.1007/978-981-10-4094-8_6

$u = 2\partial_x^2 \ln \tau$ remains exponentially small at all points in the interior of any given dominant region but is localized at the *boundaries* of two distinct regions where a balance exists between two dominant exponentials in the τ-function (6.1).

6.1.1 Tropical Approximation of the Contour Plot

To describe an asymptotic structure of the solutions, we rescale the variables (x, y, t) with a small positive number ε,

$$x \to \frac{x}{\varepsilon}, \qquad y \to \frac{y}{\varepsilon}, \qquad t \to \frac{t}{\varepsilon}.$$

In this scale, the τ-function becomes

$$\tau_A\left(\frac{x}{\varepsilon}, \frac{y}{\varepsilon}, \frac{t}{\varepsilon}\right) = \sum_{J\in\mathscr{M}(A)} \exp\left(\frac{1}{\varepsilon}\Theta_J(x, y, t) + \ln(K_J\Delta_J(A))\right)$$

$$\Theta_J(x, y, t) = \sum_{n=1}^{N}\left(\kappa_{j_n}x + \kappa_{j_n}^2 y + \kappa_{j_n}^3 t\right) = \kappa_J^{(1)}x + \kappa_J^{(2)}y + \kappa_J^{(3)}t,$$

where $\kappa_J^{(m)} = \sum_{n=1}^{N}\kappa_{j_n}^m$ for $J = \{j_1, \ldots, j_n\}$. Then we define a piecewise linear function given by the limit called a *tropical* limit or *ultra-discrete* limit [126],

$$f_{\mathscr{M}(A)}(x, y, t) := \lim_{\varepsilon\to 0}(\varepsilon \ln \tau_A) = \max_{J\in\mathscr{M}(A)}\{\Theta_J(x, y, t)\}. \tag{6.2}$$

Definition 6.1 For each t, we define the *soliton graph* of the solution $u_A(x, y, t)$ by

$$\mathscr{C}_t(\mathscr{M}(A)) = \{\text{the locus of the } xy\text{-plane where } f_{\mathscr{M}(A)}(x, y, t) \text{ is } not \text{ linear}\}.$$

Note that each region of the complement of $\mathscr{C}_t(\mathscr{M}(A))$ is a domain of linearity for $f_{\mathscr{M}(A)}(x, y, t)$, hence each region is associated to a *dominant phase* $\Theta_J(x, y, t)$ for a certain $J \in \mathscr{M}(A)$. Then the soliton graph shows the spatial pattern generated by the intersections among the dominant planes $z = \Theta_J(x, y, t)$ for $J \in \mathscr{M}(A)$.

Example 6.1 Consider $\tau_A = e^{\theta_1} + ae^{\theta_2} + be^{\theta_3}$ for Gr(1, 3). We have

$$f_{\mathscr{M}(A)}(x, y, t) = \max_{1\le j\le 3}\{\theta_j(x, y, t)\},$$

and $\mathscr{C}_t(\mathscr{M}(A))$ is given by the projection of the intersections of three planes $z = \theta_j(x, y, t)$, $j = 1, 2, 3$ onto the xy-plane, giving a Y-shape graph (see Fig. 1.3).

In this chapter, we classify the asymptotic structure of the soliton graphs $\mathcal{C}_t(\mathcal{M}(A))$ for each t in a large scale of the xy-plane. We first need to set *genericity* for the κ-parameters:

Definition 6.2 A set $\{\kappa_1, \ldots, \kappa_M\}$ is *generic* if the following sums for any m-subset of parameters $\{\kappa_{j_1}, \ldots, \kappa_{j_m}\}$ are all distinct,

$$\sum_{l=1}^{m} \kappa_{j_l}, \qquad \text{for} \quad m = 1, \ldots, \left\lfloor \frac{M+1}{2} \right\rfloor.$$

The genericity condition implies that

(1) for $m = 1$, we can assume the order $\kappa_1 < \cdots < \kappa_M$,
(2) for $m = 2$, all the solitons have *different slopes* (recall that the slope of $[i, j]$-soliton is given by $\tan \Psi_{[i,j]} = \kappa_i + \kappa_j$ which is obtained from $\theta_i = \theta_j$),
(3) for $m = 3$, all the *trivalent vertices* are separated (recall that an Y-shape soliton is given by the interaction of three planes, i.e. $\theta_i = \theta_j = \theta_l$, in particular, the y-coordinate of the trivalent vertex is $y = -(\kappa_i + \kappa_j + \kappa_l)t$).

Then we have the following useful lemma:

Proposition 6.1 *Let $\{\kappa_1 < \cdots < \kappa_M\}$ be generic. Then the dominant phases in adjacent regions of the soliton graph $\mathcal{C}_t(\mathcal{M})$ are of the form Θ_I and Θ_J with $|I \cap J| = N - 1$. That is, the index sets I and J differ by only one element.*

Proof For fixed t, let (x_0, y_0) be a point of the intersection line, $\Theta_I(x_0, y_0, t) = \Theta_J(x_0, y_0, t)$. Then using the shifted coordinates $(\bar{x} = x - x_0, \bar{y} = y - y_0)$, we have

$$\Theta_I(\bar{x}, \bar{y}, 0) = \sum_{i \in I} \theta_i(\bar{x}, \bar{y}, 0) = \sum_{j \in J} \theta_j(\bar{x}, \bar{y}, 0) = \Theta_J(\bar{x}, \bar{y}, 0).$$

Now we consider a line $\bar{x} + c\bar{y} = 0$ with a parameter c representing the slope of the line. Then along this line, each $\theta_i(\bar{x}, \bar{y}, 0)$ can be expressed by

$$\theta_i(\bar{x}, \bar{y}, 0) = \kappa_i(\kappa_i - c)\bar{y} =: \eta_i(c)\bar{y}.$$

Let c_0 be the slope of the intersection line $\Theta_I = \Theta_J$, i.e. this line is given by $\bar{x} + c_0\bar{y} = 0$. Then, without loss of generality, one can assume for $\bar{y} > 0$ that $\Theta_I > \Theta_J$ if $c < c_0$, and $\Theta_I < \Theta_J$ if $c > c_0$. That is, we have the dominant relation,

$$\begin{cases} \sum_{i \in I} \eta_i(c) > \sum_{j \in J} \eta_j(c), & \text{if } c < c_0, \\ \sum_{i \in I} \eta_i(c) < \sum_{j \in J} \eta_j(c), & \text{if } c > c_0. \end{cases}$$

Considering the set of lines $\{w = \eta_i(c) : i = 1, \dots, M\}$ in the cw-plane, one can see that any pair of lines $\{w = \eta_n(c), w = \eta_m(c)\}$ intersects at $c = \kappa_n + \kappa_m$ (cf. Fig. 1.2). Then, notice that the dominant relation can change only if some of the functions $\{\eta_i(c) : i \in I\}$ intersect at $c = c_0$ with others in $\{\eta_j(c) : j \in J\}$.

Because of the genericity condition (Definition 6.2), in particular, $\kappa_k + \kappa_l \neq \kappa_n + \kappa_m$ if $\{k, l\} \neq \{n, m\}$, there is only one pair $\{i, j\}$ for some $i \in I$ and $j \in J$ such that $\eta_i(c)$ and $\eta_j(c)$ can intersect in an interval $c_0 - \varepsilon < c < c_0 + \varepsilon$ for small positive number ε. This implies that $J = (I \setminus \{i\}) \cup \{j\}$, and completes the proof. $\qquad\square$

Proposition 6.1 implies that the asymptotic behavior of the KP solution is given by

$$u(x, y, t) \approx \tfrac{1}{2}(\kappa_j - \kappa_i)^2 \mathrm{sech}^2 \tfrac{1}{2}(\theta_j - \theta_i + \theta_{ij}) \tag{6.3}$$

in the neighborhood of the line $x + (\kappa_i + \kappa_j)y = \text{constant}$, which forms the boundary between the regions of dominant exponentials E_I and E_J with $I \setminus \{i\} = J \setminus \{j\}$. Equation (6.3) defines an *asymptotic* $[i, j]$-*soliton* as a result of those two dominant exponentials. Then the genericity condition (2) implies that those asymptotic solitons have the different slopes, such that they all separate asymptotically for $|y| \gg 0$.

We are interested in the combinatorial structure of asymptotic contour plots (*soliton graph*) which describes how line-solitons interact with each other. Generically, we expect a point at which several line-solitons meet to have degree 3. We regard such a point as a *trivalent vertex*. Three line-solitons meeting at a trivalent vertex exhibit a *resonant interaction* (this corresponds to the *balancing condition* for a tropical curve). One may also have two line-solitons which cross over each other, forming an X-shape. We call this an X-*crossing* but do not regard it as a vertex.

Remark 6.1 At $t = 0$, all the planes $z = \Theta_I(x, y, t)$ for $I \in \mathscr{M}(A)$ contain the origin of the xy-plane, hence all the lines corresponding to asymptotic solitons for $|y| \gg 0$ in $\mathscr{C}_0(\mathscr{M}(A))$ intersect at the origin (see the figure above). For $t \neq 0$, the plots are self-similar to either a plot for $t > 0$ or for $t < 0$. That is, there are essentially two soliton graphs depending only on the sign of t.

We then define the *genericity* for the soliton graphs.

Definition 6.3 A soliton graph $\mathscr{C}_t(\mathscr{M}(A))$ is *generic* if any intersection point of the plot is either trivalent vertex or X-crossing. Note that $\mathscr{C}_t(\mathscr{M}(A))$ is generic if $t \neq 0$ (see Remark 6.1).

6.2 Classification Theorem

To identify the asymptotic line-solitons for each KP soliton, we need to determine particular dominant phase combinations Θ_I in the τ-function whose indices are given by $\mathscr{M}(A)$ of matrix A representing a point of $\mathrm{Gr}(N, M)_{\geq 0}$. First we assume that A is *irreducible* as defined below.

Definition 6.4 An $N \times M$ matrix A is irreducible if

(a) each column of A contains at least one nonzero element, and
(b) each row of A contains at least one nonzero element other than the pivot once A is in RREF.

The irreducibility of A can also be stated in terms of the matroid $\mathcal{M}(A)$ as follows: A is irreducible if

(a) for any $i \in [n]$, there exists $I \in \mathcal{M}(A)$ such that $i \in I$, and
(b) there is no common index in all the elements of $\mathcal{M}(A)$, i.e. $\cap_{I \in \mathcal{M}(A)} I = \emptyset$.

Also note that the irreducibility of $A \in \mathrm{Gr}(N, M)_{\geq 0}$ is equivalent to that of the $\mathcal{J}$-diagram for the matrix A.

The reason for this assumption is the following: If an $N \times M$ matrix A is *not* irreducible, then the solution $u = 2\partial_x^2 \ln \tau$ can be obtained from a τ-function with a matrix $\tilde{A}$ of smaller size than the original matrix, i.e. the size of A is reducible. If a column of A is identically zero, then it is clear that we can re-express the functions f_i in terms of an $N \times (M - 1)$ coefficient matrix $\tilde{A}$ obtained from A by deleting its zero column. Or, suppose that an $N \times M$ matrix A in RREF has a row whose elements are all zero except for the pivot, then it can be deduced from (6.1) that the corresponding τ-function gives the same KP soliton u, which can be obtained from another τ-function associated with an $(N - 1) \times (M - 1)$ matrix $\tilde{A}$.

We now present a classification Theorem of the KP soliton solutions by identifying the asymptotic line-solitons as $y \to \pm\infty$ (see also [16, 25, 27, 71, 73]).

Theorem 6.1 *Let $A \in \mathscr{P}_{v,w}^{>0} \subset \mathrm{Gr}(N, M)_{\geq 0}$, where (v, w) defines an irreducible $\mathcal{J}$-diagram. Let $\{i_1, \ldots, i_N\}$ be the pivot set of A and $\{j_1, \ldots, j_{M-N}\}$ be the non-pivot set. Also assume that the set $\{\kappa_1 < \cdots < \kappa_M\}$ is generic.*

Then the KP soliton $u(x, y, t)$ has the following asymptotic structure (see Fig. 6.1):

(a) *For $y \gg 0$, there exists N line-solitons of $[i_n, p_n]$-types for some $p_n > i_n$. From left to right, these solitons are located in decreasing order of the slope $\kappa_{i_n} + \kappa_{p_n}$.*
(b) *For $y \ll 0$, there exists $(M - N)$ line-solitons of $[q_m, j_m]$-types for some $q_m < j_m$. From left to right, these solitons are located in increasing order of $\kappa_{q_m} + \kappa_{j_m}$.*

Moreover, the pairing map $\pi : [M] \to [M]$ defined by

$$\begin{cases} \pi(i_n) = p_n, & n = 1, \ldots, N, \\ \pi(j_m) = q_m, & m = 1, \ldots, M - N, \end{cases}$$

is a derangement given by $\pi = vw^{-1}$.

In order to prove the Theorem, we need to identify the dominant phases Θ_I with some $I \in \mathcal{M}(A)$ for the asymptotic regions for $|y| \gg 0$. To do this, consider a line $L : x + cy = 0$. Along this line, the phase function $\theta_j(x, y, t)$ takes

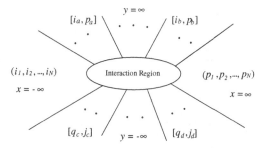

Fig. 6.1 Asymptotic structure of the KP soliton. The asymptotic line-solitons are denoted by their index pairs $[i_n, p_n]$ for $n = 1, \ldots, N$ and $[q_m, j_m]$ for $m = 1, \ldots, M - N$. The sets $(i_1, i_2, \ldots, i_N)$ and $(j_1, j_2, \ldots, j_{M-N})$ indicate pivot and non-pivot indices, respectively

$$\theta_j = \kappa_j(\kappa_j - c)\, y + \kappa_j^3 t.$$

Since we consider a large $|y|$, we ignore the term $\kappa_j^3 t$ for a fixed t. That is, we have

$$\theta_j \approx \eta_j(c)y, \quad \text{with} \quad \eta_j := \kappa_j(\kappa_j - c).$$

Recall that the $[i, j]$-soliton has the slope $\kappa_i + \kappa_j$, and we define a line parallel to this soliton by

$$L_{i,j} : \; x + (\kappa_i + \kappa_j)y = 0. \tag{6.4}$$

Before proving Theorem 6.1, we first establish the following lemma:

Lemma 6.1 *Let $\{\kappa_1 < \cdots < \kappa_M\}$ be generic (in particular, $\kappa_i + \kappa_j$ are all distinct). Then along the line $L_{i,j}$ for $i > j$, we have the dominant relation,*

$$\begin{cases} \text{(a)} \; \eta_i = \eta_j < \eta_m & \text{for} \;\; 1 \le m < i, \; \text{or} \; j < m \le n, \\ \text{(b)} \; \eta_i = \eta_j > \eta_m & \text{for} \;\; i < m < j. \end{cases}$$

Proof See the graph below for $\eta(k) = k(k - c)$ with $c = \kappa_i + \kappa_j$. This graph shows the dominant relation among the phases η_i's along the line $L_{i,j}$. □

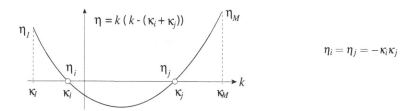

$$\eta_i = \eta_j = -\kappa_i \kappa_j$$

Because of the ordering $\kappa_1 < \cdots < \kappa_M$, one can easily see the following:

(a) For $x \ll 0$, the phase $\Theta_I = \sum_{i \in I} \theta_i$ dominates if I is the *lexicographically mini-mum* in $\mathcal{M}(A)$, i.e. I is the pivot set.
(b) For $x \gg 0$, the phase $\Theta_J = \sum_{j \in J} \theta_j$ dominates if J is the *lexicographically max-imum* in $\mathcal{M}(A)$.

Recall that these index sets I and J are given by

$$I = w \cdot \{M - N + 1, M - N + 2, \ldots, M\}, \quad J = v \cdot \{M - N + 1, M - N + 2, \ldots, M\}.$$

Here (v, w) is the pair for the Deodhar component for the matrix $A \in \mathscr{P}_{v,w}^{>0} \subset \mathrm{Gr}(N, M)_{\geq 0}$. Lemma 6.1 implies that we have the order relation among the phases Θ_I for $I \in \mathcal{M}(A)$ along the line $L_{i,j}$ for each pair of indices $\{i, j\}$. This will be essential to prove Theorem 6.1.

Proof Consider only the case for $y \gg 0$ (cases are similar, see Problem 6.1). We assume that A is in a row echelon form (REF). There are five items to prove.

(1) *If an $[i, j]$-soliton exists, then i must be a pivot*: Let Θ_I and Θ_J be two dominant phases with $I = \{i, m_2, \ldots, m_N\}$ and $J = \{j, m_2, \ldots, m_N\}$ (recall Proposition 6.1, i.e. $|I \cap J| = N - 1$). Assume that the i-th column of the A-matrix, say A_i, is not a pivot column. Then A_i can be expanded by the pivot columns A_{i_r} for $i_r < i$, i.e.
$$A_i = \sum_{r=1}^{s} c_r A_{i_r} \text{ for } c_r \neq 0 \text{ and some } i_r < i, \text{ which leads to}$$

$$\Delta_I = \sum_{r=1}^{s} c_r \Delta_{(I \setminus \{i\}) \cup \{i_r\}}.$$

This implies that $\Delta_{(I \setminus \{i\}) \cup \{i_r\}} \neq 0$ for some $i_r < i$. But this contradicts that Θ_I is the dominant phase, since $\theta_{i_r} > \theta_i$ from Lemma 6.1.

(2) *Existence of the $[i, p]$-soliton for each pivot i*: For each i, consider $j > i$. Let j^* be the maximum j such that
$$\begin{cases} \mathrm{rank}[1, \ldots, i - 1, j, j + 1, \ldots, M] = N, \\ \mathrm{rank}[1, \ldots, i - 1, j + 1, \ldots, M] = N - 1, \end{cases}$$

where $[k_1, k_2, \ldots, k_n]$ represents the submatrix of A obtained from columns with the indices $k_1, \ldots, k_n$. Then we can see from Lemma 6.1 that there are two following cases to find p for the $[i, p]$-soliton:

(a) If $\mathrm{rank}[1, \ldots, i, j^* + 1, \ldots, M] = N$, then take $p = j^*$. That is, choose $\{m_2, \ldots, m_N\} \subset \{1, \ldots, i - 1, j^* + 1, \ldots, M\}$, such that rank $[m_2, \ldots, m_N] = N - 1$, and Θ_I and Θ_J with $I = \{i, m_2, \ldots, m_N\}$ and $J = (I \setminus \{i\}) \cup \{p\}$ are two dominant phases along the line $L_{i,p} : x + (\kappa_i +$

$\kappa_p)y = 0$. That is, we have $I, J \in \mathcal{M}(A)$ and the ordering $\Theta_I = \Theta_J > \Theta_L$ for all $L \in \mathcal{M}(A) \setminus \{I, J\}$ along $L_{i,p}$.

(b) If $\text{rank}[1, \ldots, i, j^* + 1, \ldots, M] = N - 1$, then $A_i \in \text{Span}[1, \ldots, i - 1, j^* + 1, \ldots, M]$. Let $r^* > i - 1$ be the maximum r such that

$$\begin{cases} A_i \in \text{Span}[1, \ldots, i - 1, r, r + 1 \ldots, M], \\ A_i \notin \text{Span}[1, \ldots, i - 1, r + 1, \ldots, M]. \end{cases}$$

Then take $p = r^*$ and choose linear independent sets $\{m_1, \ldots, m_k, m_{k+1}, \ldots, m_N\}$ such that

$$\begin{cases} \text{Span}[m_2, \ldots, m_k] = \text{Span}[1, \ldots, i - 1, r^* + 1 \ldots, M], \\ \text{Span}[m_{k+1}, \ldots, m_N] = \text{Span}[i + 1, \ldots, r^* - 1], \end{cases}$$

and Θ_I and Θ_J with $I = \{i, m_2, \ldots, m_N\}$ and $J = (I \setminus \{i\}) \cup \{p\}$ are two dominant phases along the line $L_{i,p}$.

(3) *Uniqueness of the $[i, p]$-soliton*: Suppose that there are two such indices p and p'. Assume $p > p'$, then the $[i, p]$-soliton appears to the left of the $[i, p']$-soliton (due to the slope relation). Recall that the index set in the dominant phase in the region for $x \ll 0$ consists of the pivots, $\{i_1, \ldots, i_N\}$. Since the pivot index i is replaced by $[i, p]$-soliton, there must exist an $[i', i]$-soliton with a pivot $i' < i$ between $[i, p]$- and $[i', i]$-solitons. However this is impossible, since

$$i' < i < p' \quad \text{implies} \quad \kappa_i + \kappa_{p'} > \kappa_{i'} + \kappa_i,$$

i.e. the $[i', i]$-soliton is on the right of $[i, p']$-soliton (not on the left of).

(4) *The pairing map π is a derangement*: We consider the phase function θ_j for KP hierarchy, i.e. we have

$$\theta_j = \sum_{a=1}^{\infty} \kappa_j^a t_a, \quad j = 1, \ldots, M.$$

Since $u = 2\partial_x^2 \ln \tau$, the t_a-flow of the KP hierarchy can be written in the form (see (2.19))

$$\partial_{t_a} u = \partial_x F_a,$$

where F_a is given by $F_a = 2\partial_x \partial_{t_a} \ln \tau$. Note also that (2.19) gives a conservation law for the conserved density F_a,

$$\partial_{t_b} F_a = \partial_x G_{a,b}, \quad \text{with} \quad G_{a,b} = 2\partial_{t_a} \partial_{t_b} \ln \tau.$$

Then we have the conserved quantity C_a by taking the integral of F_a,

$$C_a := \frac{1}{2} \int_{x \ll 0}^{x \gg 0} F_a \, dx = \partial_{t_a} \ln \tau \Big|_{x \gg 0} - \partial_{t_a} \ln \tau \Big|_{x \ll 0}.$$

For $y \gg 0$ (resp. $y \ll 0$), we have shown that there are N solitons of $[i_n, p_n]$-type (resp. there are $M - N$ solitons of $[q_m, j_m]$-type from Problem 6.1) the integral C_a becomes

$$C_a^+ = \sum_{n=1}^{N} (\kappa_{p_n}^a - \kappa_{i_n}^a) \quad \left(\text{resp. } C_a^- = \sum_{m=1}^{M-N} (\kappa_{q_m}^a - \kappa_{j_m}^a) \right).$$

Noting that C_a does not depend on any t_b, and in particular, for $t_b = y$, we have $C_a^+ = C_a^-$, which leads to

$$\sum_{i=1}^{M} \kappa_i^a = \sum_{n=1}^{N} \kappa_{i_n}^a + \sum_{m=1}^{M-N} \kappa_{j_m}^a = \sum_{n=1}^{N} \kappa_{p_n}^a + \sum_{m=1}^{M-N} \kappa_{q_m}^a.$$

These equations (power sums) for $a = 1, \ldots, M$ imply that

$$\{p_1, \ldots, p_N, q_1, \ldots, q_{M-N}\} = \{1, 2, \ldots, M\},$$

that is, π is a derangement.

(5) *The pairing map π is given by $\pi = vw^{-1}$.* Here, we prove the case $\pi(i_n) = p_n$ for $n = 1, \ldots, N$ (other case where $\pi(j_m) = q_m$ for $y \ll 0$ is similar). First note from Lemma 6.1 that the set of pivot indices $I_1 := \{i_1, \ldots, i_N\}$ gives the dominant phase Θ_{I_1} for $x \ll 0$. From (1) above, there exists a line soliton of type $[i_n, p_n]$, which has the slope $c_n = \kappa_{i_n} + \kappa_{p_n}$. For simplicity, we assume that the slopes of the asymptotic solitons in $y \gg 0$ are ordered as

$$\kappa_{i_1} + \kappa_{p_1} < \kappa_{i_2} + \kappa_{p_2} < \cdots < \kappa_{i_N} + \kappa_{p_N}.$$

(The order depends on the choice of the κ-parameters and the following argument can still be applied for the general case.) Then the leftmost soliton is of $[i_N, p_N]$-type and appears at the boundary between the regions of the dominant phases Θ_{I_1} and Θ_{I_2} with $I_2 = (I_1 \setminus \{i_N\}) \cup \{p_N\}$. From (1), one can see that the index set I_2 is the lexicographically maximal element of the index set

$$\mathcal{M}(A) \cap \{\{i_1, \ldots, i_{N-1}, q\} : i_N < q \leq M\} \quad \text{for fixed } \{i_1, \ldots, i_{N-1}\},$$

and $p_N = vw^{-1}(i_N) = \pi(i_N)$. After having $p_l = \pi(i_l)$ for $l = k + 1, \ldots, N$, one can see that there is an $[i_k, p_k]$-soliton at the boundary of the region, where Θ_{I_k} with I_k is the lexicographically maximal element of

$$\mathcal{M}(A) \cap \{\{i_1, \ldots, i_{k-1}, q_k, p_{k+1} \ldots, p_M\} : i_k < q_k \leq M\}.$$

Then the lexicographically maximal element of this set is given by

$$\{i_1, \ldots, i_{k-1}, p_k, \ldots, p_M\} \quad \text{with} \quad p_k = \pi(i_k).$$

Continuing the process, we have $\pi(i_k) = p_k$ for $k = 1, \ldots, N$. Then proving the cases for the non-pivot indices (see Problem 7.2), we conclude that the pairing map π is given by $\pi = vw^{-1}$.

The items (1) through (5) above complete the proof. □

The following example illustrates how one can find the asymptotic solitons using the procedure given in the proof of Theorem 6.1.

Example 6.2 Consider the 2×2 matrix

$$A = \begin{pmatrix} 1 & 0 & 0 & -c \\ 0 & 1 & a & b \end{pmatrix} \quad \text{with} \quad a, b, c > 0.$$

This matrix is an element in $\mathscr{P}_{v,w} \in \mathrm{Gr}(2, 4)_{\geq 0}$, where $w = s_2 s_3 s_1 s_2$ and $v = s_1$, i.e. $\pi = (3, 4, 2, 1)$. The matroid $\mathscr{M}(A)$ is then given by

$$\mathscr{M}(A) = \{12, 13, 14, 24, 34\},$$

i.e. the minor $\Delta_{23}(A) = 0$. According to Theorem 6.1, there are two asymptotic line-solitons for both $y \gg 0$ and $y \ll 0$. They are identified by the index pairs $[1, p_1]$ and $[2, p_2]$ for $y \gg 0$, and $[q_1, 3]$ and $[q_2, 4]$ for $y \ll 0$. We first determine the asymptotic soliton for $y \gg 0$ using the rank conditions described in the item (2) of the proof. For the first pivot column $i_1 = 1$, the maximum index j, such that $\mathrm{rank}[j, 4] = 2$ and $\mathrm{rank}[j + 1, 4] = 1$, is given by $j^* = 3$, i.e. $\{3, 4\} \in \mathscr{M}(A)$. Since $\theta_4 > \theta_1 = \theta_3 > \theta_2$ along the line $L_{1,3} : x + (\kappa_1 + \kappa_3)y = 0$ (Lemma 6.1), the two dominant phases are given by $\Theta_{3,4}$ and $\Theta_{1,4}$. Thus, the first asymptotic soliton for $y \gg 0$ is identified as a $[1, 3]$-soliton. For $i_2 = 2$, proceeding in a similar manner as above, we find that $j^* = 4$, i.e. $\{1, 4\} \in \mathscr{M}(A)$. This implies that the second asymptotic soliton is a $[2, 4]$-soliton.

Now, for $y \ll 0$, we first look for the asymptotic soliton of $[q_1, 3]$-type. Since we need to find the least phase combinations for this region, we compare the phases $\Theta_{i,j}$ with $\{i, j\} \in \mathscr{M}(A)$ along the line $L_{q_1,3}$. In the rank condition similar to the case for $y \gg 0$, one can see that at $q_1 = 2$, either $\{\Theta_{1,2}, \Theta_{1,3}\}$ or $\{\Theta_{2,4}, \Theta_{3,4}\}$ gives the pair of least phases. In either case, we have $[2, 3]$-soliton as the asymptotic soliton of $[q_1, 3]$-type. The second asymptotic soliton of $[q_2, 4]$-type can be found in a similar way. Figure 6.2 illustrates the contour plots of the KP soliton generated by the matrix A with $a = b = c = 1$ and the κ-parameters $(\kappa_1, \ldots, \kappa_4) = (-1, -0.5, 0.5, 2)$.

In Chap. 8, we will study the structure of the two-dimensional patterns generated by the KP soliton for $|t| \gg 0$.

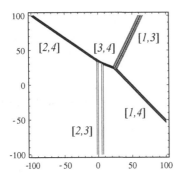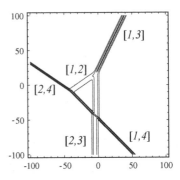

Fig. 6.2 KP soliton generated by a matrix A in $\mathscr{P}_{v,w}^{>0} \subset \mathrm{Gr}(2,4)_{\geq 0}$ with $\pi = vw^{-1} = (3,4,1,2)$

According to Theorem 6.1, the KP soliton generated by an $N \times M$ irreducible matrix A has N line-solitons for $y \gg 0$ and $M - N$ line-solitons for $y \ll 0$, which we refer to as the soliton solution (N_+, N_-)-soliton with $N_+ = N$ and $N_- = M - N$.

Definition 6.5 Let $\mathscr{S}_+ := \{[i_n, p_n] : n = 1, \dots, N\}$ and $\mathscr{S}_- := \{[q_m, j_m] : m = 1, \dots, M - N\}$ denote the index sets identifying the line-solitons for $y \gg 0$ and for $y \ll 0$, respectively. Then two (N_+, N_-)-soliton solutions are said to be in the same equivalence class if their asymptotic line-solitons are labeled by identical sets $\mathscr{S}_\pm$ of index pairs. Note that the set $\mathscr{S} := \mathscr{S}_+ \cup \mathscr{S}_-$ is uniquely determined by the permutation $\pi = vw^{-1}$ and is denoted by $\mathscr{S} = \mathscr{S}_\pi$. These solutions in an equivalent class come from the same Deodhar component $\mathscr{P}_{v,w}^{>0} \subset \mathrm{Gr}(N, M)_{\geq 0}$.

Then there are M unbounded (polyhedral) regions in the xy-plane and a boundary of each region is given by an asymptotic line-soliton of either $[i_n, p_n]$- or $[q_m, j_m]$-type. We define a *soliton necklace* identifying the dominant phases Θ_I of those regions.

Definition 6.6 A *soliton necklace* of $\mathscr{M}(A)$ is a sequence of N-index subsets of $\mathscr{M}(A)$ defined by

$$\mathscr{I}_S := \{I_1 < I_2, < \cdots < I_N < I_{N+1} > \cdots > I_M > I_{M+1} = I_1\}.$$

Here the sequence $(I_1, \dots, I_{N+1})$ represents the set of dominant phases Θ_{I_n} from left to right for $y \gg 0$, while $(I_{N+1}, \dots, I_{M+1})$ represents the dominant phases Θ_{I_m} for $y \ll 0$ from right to left. That is, the index sets in the sequence are in clockwise order in the xy-plane according to the slope $\kappa_i + \kappa_{\pi(i)}$ of the asymptotic solitons. Note in particular that $I_1 = \{i_1, \dots, i_N\}$ is the pivot set appearing at $x \ll 0$ and $I_{N+1} = \{p_1, \dots, p_N\}$ with $p_n = \pi(i_n)$ appearing at $x \gg 0$ (see Fig. 6.1).

Remark 6.2 Since the soliton necklace depends on the choice of the κ-parameters, the soliton necklace is, in general, not the same as the *Grassmann necklace* defined in [103] (see Problem 5.3). Here the (irreducible) Grassmann necklace is defined as a sequence $\mathscr{I}_G = (I_1, \ldots, I_M)$ of $I_n \in \mathcal{M}(A)$ such that for $i \in [M]$, $I_{i+1} = (I_i \setminus \{i\}) \cup \{j\}$ for some $j \neq i$ (the indices i are taken modulo M).

Example 6.3 For the Example 5.8, a soliton necklace is given by

$$\mathscr{I}_S = \{1247 < 1248 < 1289 < 1489 < 4789 > 4578 > 4567 > 3457 > 1347 > 1247\},$$

when the clockwise order of solitons from $x \ll 0$ are given by

$$\{[7, 8], [4, 9], [2, 4], [1, 7]; [5, 9], [6, 8], [3, 6], [1, 5], [2, 3]\},$$

where the first four solitons are in $y \ll 0$ and the next five solitons are in $y \ll 0$. That is, the order of the slopes of line-solitons is given by

$$\kappa_7 + \kappa_8 > \kappa_4 + \kappa_9 > \kappa_2 + \kappa_4 > \kappa_1 + \kappa_7 \quad \text{and}$$
$$\kappa_5 + \kappa_9 > \kappa_6 + \kappa_8 > \kappa_3 + \kappa_6 > \kappa_1 + \kappa_5 > \kappa_2 + \kappa_3.$$

Note that the soliton necklace $\mathscr{I}_S$ is the same as the Grassmann necklace $\mathscr{I}_G$.

With a different choice of the κ-parameters, one can have an alternative soliton necklace

$$\mathscr{I}_S = \{1247 < 1248 < 1289 < \mathbf{2789} < 4789 > \mathbf{4679} > 4567 > 3457 > \mathbf{2457} > 1247\}.$$

Here the bold face numbers are not included in the first necklace. In the second necklace, the soliton order is given by

$$\{[7, 8], [4, 9], [1, 7], [2, 4]; [6, 8], [5, 9], [3, 6], [2, 3], [1, 5]\},$$

i.e. we have the following order of the κ-parameters,

$$\kappa_7 + \kappa_8 > \kappa_4 + \kappa_9 > \kappa_1 + \kappa_7 > \kappa_2 + \kappa_4 \quad \text{and}$$
$$\kappa_6 + \kappa_8 > \kappa_5 + \kappa_9 > \kappa_3 + \kappa_6 > \kappa_2 + \kappa_3 > \kappa_1 + \kappa_5.$$

Problems

6.1 Complete the proof of Theorem 6.1 by proving the case for $y \ll 0$ (see also Chap. 7).

6.2 Determine the asymptotic structures of KP solitons generated by the irreducible 2×4 matrices A with a generic choice of κ-parameters (see also [27]).

6.3 Let A be an $N \times M$ matrix from $Gr(N, M)_{>0}$, i.e. $\mathcal{M}(A) = \binom{[M]}{N}$. Then using Lemma 6.1, show that the KP soliton generated by A has the asymptotic line-solitons of $[i, M - N + i]$-types $(i = 1, \ldots, N)$ for $y \gg 0$ and of $[j - N, j]$-types $(j = N + 1, \ldots, M)$ for $y \ll 0$. Confirm that the pairing map π is

$$\pi = (M - N + 1, M - N + 2, \ldots, M - 1, M, N + 1, N + 2, \ldots, M - N - 1, M - N).$$

(See also [17].)

6.4 Find an example of the soliton necklace, that is different from the Grassmann necklace.

6.5 Given a set of asymptotic solitons in $|y| \gg 0$, construct an algorithm that generates a KP soliton with the same set of asymptotic solitons (see also [29]).

6.6 Consider a set of M points $\{(p_i, q_i) : i = 1, \ldots, M\}$, where $p_i = -\sin \psi_i$ and $q_i = \cos \psi_i$ for $0 \le \psi < 2\pi$ (see Sect. 3.2.3). Let A be an element of $Gr(N, M)_{\ge 0}$, and consider the matroid $\mathcal{M}(A)$. Choose $\{(p_i, q_i)\}$ so that the convex hull of points $\{(p_I, q_I) : I \in \mathcal{M}(A)\}$ with $p_I = \sum_{i \in I} p_i$ and $q_I = \sum_{i \in I} q_i$ is an M-gon.

Show that the dual graph of this M-gon can provide the set of asymptotic solitons of the Davey-Stewartson (DS) soliton generated by the τ-function given in Sect. 3.2.3. Explain how to identify the component $\mathcal{P}_{v,w}^{>0}$ for the matrix A from the DS soliton, i.e. identify the permutation $\pi = vw^{-1} \in S_M$ from M asymptotic DS line-solitons.

Chapter 7
KP Solitons on $\mathrm{Gr}(N, 2N)_{\geq 0}$

Abstract In this chapter, we will discuss the reflective symmetry $(x, y, t) \leftrightarrow (-x, -y, -t)$ of the KP equation, and show that this symmetry is a consequence of the *duality* between the Grassmannian $\mathrm{Gr}(N, M)$ and $\mathrm{Gr}(M - N, M)$ in terms of the KP solitons. Using this duality, we construct the KP solitons for $\mathrm{Gr}(M - N, M)_{\geq 0}$ from those for $\mathrm{Gr}(N, M)_{\geq 0}$. We then consider a special class of KP solitons for $\mathrm{Gr}(N, 2N)_{\geq 0}$, which consists of the same set of the asymptotic solitons at both $y \ll 0$ and $y \gg 0$, i.e. $\mathscr{S}_+ = \mathscr{S}_-$. The soliton solutions of this type are referred to as N-soliton solutions. The simplest ones of these solutions consist of N line-solitons having non-resonant interactions among those solitons and are generated from the points in an irreducible component of the *lowest* dimension, N, of $\mathrm{Gr}(N, 2N)_{\geq 0}$. We also discuss some combinatorial properties of those solutions. For the simplest cases of non-resonant interactions, the total number of such N-soliton solutions is given by a Catalan number $C_N = \frac{1}{N+1} \binom{2N}{N}$.

7.1 Duality Between $\mathrm{Gr}(N, M)_{\geq 0}$ and $\mathrm{Gr}(M - N, M)_{\geq 0}$

The KP equation (1.2) is invariant under the reflection $(x, y, t) \to (-x, -y, -t)$. Consequently, if $u(x, y, t)$ has the asymptotic line-solitons with $\mathscr{S}_+$ (cf. Definition 6.5), then $u(-x, -y, -t)$ has the asymptotic line-solitons with reversed sets $\mathscr{S}_\mp$. In terms of the permutation π, if the sets $\mathscr{S}_\pm$ are determined by π, then $\mathscr{S}_\mp$ are determined by π^{-1}. We refer to the KP solitons $u(x, y, t)$ and $u(-x, -y, -t)$, as well as their respective equivalence classes as *dual* to each other. In terms of the Grassmannian, this duality is due to the isomorphism between $\mathrm{Gr}(N, M)$ and $\mathrm{Gr}(M - N, M)$.

Let $\tau_{N,M}(x, y, t)$ denote the τ-function in (6.1) for $u(x, y, t)$, i.e.

$$\tau_{N,M} = \sum_{I \in \binom{[M]}{N}} \Delta_I(A) K_I e^{\Theta_I},$$

where $A \in \mathrm{Gr}(N, M)_{\geq 0}$ and $K_I = \prod_{l > m}(\kappa_{i_l} - \kappa_{i_m})$ for $I = \{i_1 < \cdots < i_N\}$. Then the solution $u(-x, -y, -t)$ will be generated by $\tau_{NM}(-x, -y, -t)$. Note, however, that $\tau_{N,M}(-x, -y, -t)$ does not have the form given by (6.1), and is indeed possible

© The Author(s) 2017

Y. Kodama, *KP Solitons and the Grassmannians*,
SpringerBriefs in Mathematical Physics 22, DOI 10.1007/978-981-10-4094-8_7

to construct a certain τ-function $\tau_{M-N,M}(x, y, t)$ from $\tau_{N,M}(-x, -y, -t)$ that is *dual* to $\tau_{N,M}(x, y, t)$. We describe below how to construct the function $\tau_{M-N,M}(x, y, t)$ from $\tau_{N,M}(x, y, t)$ in the form,

$$\tau_{M-N,M} = \sum_{J \in \binom{[M]}{M-N}} \Delta_J(B) K_J e^{\Theta_J},$$

with a matrix $B \in Gr(M - N, M)_{\geq 0}$.

First we obtain the coefficient matrix B for the τ-function $\tau_{M-N,M}$ from the $N \times M$ matrix A associated with $\tau_{N,M}(x, y, t)$. Since A is of full rank, its rows form a basis for an N-dimensional subspace V of $\mathbb{R}^M$. Let $V^\perp$ be the orthogonal complement of V with respect to the standard inner product on $\mathbb{R}^M$, and let A' be an $(M - N) \times M$ matrix whose rows form a basis for $V^\perp$. Clearly, A' is not unique, but a particular choice for A' is as follows: Suppose the pivot and non-pivot columns of A in RREF are represented by the $N \times N$ identity matrix I_N and the $N \times (M - N)$ matrix G, respectively. Then one can obtain such matrix A' from A,

$$A = [I_N, G] P \qquad \Longrightarrow \qquad A' = [-G^T, I_{M-N}] P, \qquad (7.1)$$

where G^T is the matrix transpose of G and P is an $M \times M$ permutation matrix satisfying $P^T = P^{-1}$. It can be directly verified from (7.1) that $AA'^T = 0$ showing the $N \times (M - N)$ orthogonality relations among the row vectors of A and A'. Now consider the $M \times M$ matrix H defined by

$$H := \begin{pmatrix} I_N & G \\ -G^T & I_{M-N} \end{pmatrix} P.$$

Then using the Laplace expansion for H, one can find

$$\det(H) = \sum_{I \cup J = [M]} \sigma(I, J) \Delta_I(A) \Delta_J(A') = \sum_{I \in \binom{[M]}{N}} (\Delta_I(A))^2,$$

where $I = \{i_1 < i_2 < \cdots < i_N\}$, $J = [M] \setminus I = \{j_1 < j_2 < \cdots < j_{M-N}\}$ and $\sigma(I, J)$ is the sign-function for the permutation $\pi = (I, J)$ (explicitly it can be given by $\sigma(I, J) = (-1)^{\frac{N(N+1)}{2} + i_1 + \cdots + i_N}$). This implies that we have

$$\Delta_J(A') = \sigma(I, J) \Delta_I(A).$$

Hence, if the Plücker coordinate $\Delta_I(A)$ is nonzero, then the same holds for the corresponding coordinate $\Delta_J(A')$. The sign factor $\sigma(I, J) = \pm 1$ can be rescaled by an $(M - N) \times (M - N)$ orthogonal matrix O and the $M \times M$ diagonal matrix $D = \mathrm{diag}(-1, 1, -1, \ldots, (-1)^M)$ such that

$$\Delta_I(A) = \Delta_J(B) \quad \text{where} \quad B = OA'D. \tag{7.2}$$

The $(M - N) \times M$ matrix B plays the role of a coefficient matrix for the τ-function $\tau_{M-N,M}(x, y, t)$, which is related to the function $\tau_{N,M}(-x, -y, -t)$. Indeed, under the transformation $(x, y, t) \to (-x, -y, -t)$, we have

$$\tau_{N,M}(-x, -y, -t) = \exp[-\Theta_{[M]}(x, y, t)] \, \tau'(x, y, t),$$

where $\Theta_{[M]}(x, y, t) = \sum_{m=1}^{M} \theta_m(x, y, t)$ with $\theta_m(x, y, t) = \kappa_m x + \kappa_m^2 y + \kappa_m^3 t + \theta_m^0$. Here the function $\tau'(x, y, t)$ is given by

$$\tau'(x, y, t) = \sum_{J \in \binom{[M]}{M-N}} K_I \Delta_J(B) \exp[\Theta_J(x, y, t)], \tag{7.3}$$

where $I = [M] \setminus J$. Now, we take the constants θ_m^0 in the phase function $\theta_m(x, y, t)$ by

$$\theta_m^0 = \sum_{r \neq m} \ln |\kappa_r - \kappa_m|, \quad m = 1, 2, \ldots, M,$$

which satisfy the identity

$$\exp \Theta_J^0 = \prod_{m=1}^{M-N} \prod_{r \neq j_m} |\kappa_r - \kappa_{j_m}| = K_J K_I^{-1} K_{[M]} \quad \text{with} \quad I \cup J = [M],$$

where $\Theta_J^0 = \sum_{m=1}^{M-N} \theta_{j_m}^0$ and $K_{[M]} = \prod_{i<j}(\kappa_j - \kappa_i)$. Then we obtain the dual τ-function

$$\tau_{M-N,M}(x, y, t) = K_{[M]}^{-1} \tau'(x, y, t) = \sum_{J \in \binom{[M]}{M-N}} \Delta_J(B) K_J \exp[\Theta_J(x, y, t)]. \tag{7.4}$$

We summarize the result as the following Proposition:

Proposition 7.1 *Let $u(x, y, t)$ be a KP soliton on $A \in$ Gr(N, M)$_{\geq 0}$. Let B be the $(M - N) \times (M - N)$ matrix defined by (7.2), i.e. $B \in$ Gr($M - N$, M)$_{\geq 0}$, the dual Grassmannian. Then we have the following:*

(a) *The solution obtained by $u(-x, -y, -t)$ is a KP soliton on $B \in$ Gr($M - N$, M)$_{\geq 0}$.*
(b) *If $I \in \binom{[M]}{N}$ and $J = [M] \setminus I$, then $\Delta_I(A) = \Delta_J(B)$.*
(c) *If $\pi \in S_M$ is the pairing map for $u(x, y, t)$, then the pairing map for the solution $u(-x, -y, -t)$ is given by π^{-1}.*

7.1.1 Duality Between the Le-Diagrams

We here demonstrate how one can compute the dual matrix $B \in \mathrm{Gr}(M - N, M)_{\geq 0}$ from a given matrix $A \in \mathrm{Gr}(N, M)_{\geq 0}$ using the conjugate relation between the corresponding $\mathcal{J}$-diagrams.

Let $L(A)$ be the $\mathcal{J}$-diagram associated to the matrix $A \in \mathscr{P}_{v,w}^{>0}$ with $w \in S_M^{(N)}$ and a PDE $v \leq w$. Then the conjugate diagram of $L(A)$ is defined by flipping the diagram over its diagonal line (see below). The conjugate digram then gives the $\mathcal{J}$-diagram for the matrix $B \in \mathrm{Gr}(M - N, M)$, which we denote $L(B)$. The reading order of the conjugate diagram $L(B)$ is the same as that of $L(A)$, but the numbering in the southeast boundary of the diagram $L(B)$ is now in the counterclockwise direction, i.e. starting from the bottom-left corner with the label 1 and ending at the top-right corner with the label M.

Notice that the expression of w can be considered as either $w \in S_M^{(N)}$ or $w \in S_M^{(M-N)}$. For example, one can see that the following two expressions for the pair (v, w), corresponding to the diagrams above, are the same, i.e.

$$w = s_7 \bar{s}_8 s_4 s_5 s_6 \bar{s}_7 s_3 \bar{s}_4 \bar{s}_5 \bar{s}_6 s_1 \bar{s}_2 s_3 \bar{s}_4 s_5 = s_1 s_4 s_3 \bar{s}_2 s_5 \bar{s}_4 s_3 s_7 s_6 \bar{s}_5 \bar{s}_4 s_8 \bar{s}_7 \bar{s}_6 s_5$$

Here, $\bar{s}_j$ indicates the word from the subexpression $v = s_8 s_7 s_4 s_5 s_6 s_2 s_4 = s_2 s_4 s_5 s_4 s_8 s_7 s_6$. The chord diagrams corresponding to these two equivalent expressions are shown in the figures below.

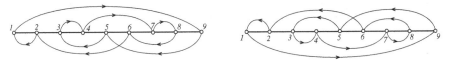

The conjugation in the $\mathcal{J}$-diagrams corresponds to flipping the chord upside down and the pivot set for $L(A)$ becomes the non-pivot set for $L(B)$. Note that the pivot set $\{1, 3, 4, 7\}$ appears at the east boundary of $L(A)$, and the non-pivot set $\{2, 5, 6, 8, 9\}$ appears at the east boundary of $L(B)$. The arrows in the chord diagrams indicate the exchange of the pivot-nonpivot relationship.

The matrices A and B can be found from the corresponding networks below.

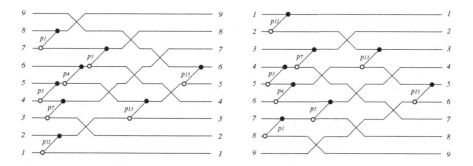

The matrix A is calculated as

$$A = \begin{pmatrix} p_{11}p_{13}p_{15} & p_{13}p_{15} & 0 & -p_3 p_{15} & -p_{15} & 0 & 0 & 0 & -1 \\ 0 & 0 & p_7 & 1 & 0 & 0 & 0 & 0 & 0 \\ 0 & 0 & 0 & p_3 p_4 p_5 & p_4 p_5 & p_5 & 1 & 0 & 0 \\ 0 & 0 & 0 & 0 & 0 & 0 & p_1 & 1 & 0 \end{pmatrix}$$

and the matrix B is given by

$$B = \begin{pmatrix} 1 & p_{11} & 0 & 0 & 0 & 0 & 0 & 0 & 0 \\ 0 & 0 & 1 & p_7 & p_3 p_7 & 0 & 0 & 0 & 0 \\ 0 & -1 & 0 & 0 & p_{13} & p_4 p_{13} & 0 & 0 & 0 \\ 0 & 0 & 0 & 0 & 0 & 1 & p_5 & p_1 p_5 & 0 \\ 0 & 0 & 0 & 0 & -1 & -p_4 & 0 & 0 & p_{15} \end{pmatrix}.$$

Note here that the labeling of the network for B (right network) is opposite from the network for A (left network). Note also that the lexicographically maximum element in $\mathcal{M}(B)$ is given by $\{2, 5, 6, 8, 9\}$, which is the non-pivot set of A. Each non-pivot index, say j_k, appears at the last nonzero element of the k-th row, and the first nonzero element of this row is given by $\pi(j_k)$ with $\pi = vw^{-1}$. The B matrix is given by the first five columns of the matrix $g \in G_{v,w}^{>0}$ generated from the right network, i.e.

$$g = \begin{pmatrix} g_{1,1} & \cdots & g_{1,M-N} & \cdots & g_{1,M} \\ \vdots & & \vdots & & \vdots \\ g_{M,1} & \cdots & g_{M,M-N} & \cdots & g_{M,M} \end{pmatrix} \mapsto B = \begin{pmatrix} g_{1,1} & \cdots & g_{M,1} \\ \vdots & & \vdots \\ g_{1,M-N} & \cdots & g_{M,M-N} \end{pmatrix}.$$

Remark 7.1 One can use the dual diagram to prove Theorem 6.1 for the case with $y \ll 0$ (see Problem 6.1). That is, under the change of the coordinates $(x, y, t) \to (-x, -y, -t)$, the argument used in the proof can be applied for this case. Then the pivot set for the B matrix is now identified as the non-pivot set for the A matrix (see the dual J-diagram above) and is given by the lexicographically maximal element of $\mathcal{M}(B)$, which is $J = w\{1, 2, \ldots, M - N\}$.

7.2   The N-Soliton Solutions

When $M = 2N$, we have the (N, N)-soliton solutions, i.e. the numbers of asymptotic line-solitons as $y \to \infty$ and as $y \to -\infty$ are the same. However, in general, the sets $\mathscr{S}_+$ and $\mathscr{S}_-$ are different. A particularly interesting subclass of the (N, N)-solitons consists of KP solitons, which have identical sets of asymptotic line-solitons as $|y| \to \infty$, i.e., $\mathscr{S}_- = \mathscr{S}_+$. Each solution of this class will be referred to as N-soliton solution. The main features of these solutions are listed below.

Property 7.1 *The N-soliton solutions satisfy the following properties.*

1. *The τ-function of an N-soliton solution, denoted by τ_N, is expressed in terms of $2N$ distinct κ-parameters and an $N \times 2N$ matrix A, which is irreducible and its Plücker coordinates satisfy the duality conditions due to (7.2), i.e. for any pair of N-index subsets (I, J) with $I \cup J = [M]$,*

$$\Delta_I(A) = 0 \quad \text{if and only if} \quad \Delta_J(A) = 0. \tag{7.5}$$

   That is, $I \in \mathcal{M}(A)$ if and only if $J \in \mathcal{M}(A)$. In terms of τ_N, this implies that the phase combination $\Theta_I(x, y, t)$ is present in τ_N if and only if $\Theta_J(x, y, t)$ is also present.
2. *Each N-soliton solution has exactly N asymptotic line-solitons as $y \to \pm\infty$ identified by the same index pairs $[i_n, j_n]$ with $i_n < j_n$, $n = 1, \ldots, N$. The sets $\{i_1, \ldots, i_N\}$ and $\{j_1, \ldots, j_N\}$ label the pivot and non-pivot columns of the coefficient matrix A, respectively.*
3. *The pairing maps associated with N-soliton solutions are involutions of S_{2N} with no fixed points, defined by the set $I_{2N} = \{\pi \in S_{2N} | \pi^{-1} = \pi, \pi(i) \neq i, \forall i \in [2N]\}$. Such permutations can be expressed as products of N disjoint 2-cycles, and their chord diagrams are self-dual, i.e. symmetric about the horizontal axis (see Sect. 7.3.1). The total number of such involutions is given by $|I_{2N}| = (2N - 1)!! = 1 \cdot 3 \ldots \cdot (2N - 1)$ [18] (see also [69]). Hence, there are $(2N - 1)!!$ distinct equivalence classes of N-soliton solutions.*

Examples of the N-soliton solutions with special choices of the functions $\{f_n : n = 1, \ldots, N\}$ in (3.6) and the $N \times 2N$ matrix A are given below.

Example 7.1 O-type N-soliton solutions: These are the well-known [43, 56, 115] multi-soliton solutions of KP constructed by choosing $\{f_n : n = 1, \ldots, N\}$ according to

$$f_n(x, y, t) = e^{\theta_{2n-1}} + a_n e^{\theta_{2n}}, \quad n = 1, \ldots, N.$$

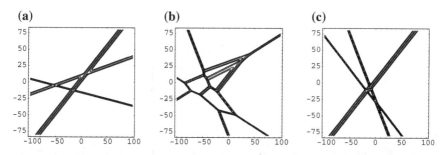

Fig. 7.1 Three different 3-soliton solutions with the same phase parameters $(\kappa_1, \ldots, \kappa_6) = (-3, -2, 0, 1, -\frac{3}{2}, 2)$, illustrating the three classes: (**a**) O-type, (**b**) T-type, and (**c**) P-type 3-soliton solutions

The corresponding matrix A is given by

$$A_O = \begin{pmatrix} 1 & a_1 & 0 & 0 & \cdots & 0 & 0 \\ 0 & 0 & 1 & a_2 & \cdots & 0 & 0 \\ \vdots & \vdots & \vdots & \vdots & \ddots & \vdots & \vdots \\ 0 & 0 & 0 & 0 & \cdots & 1 & a_N \end{pmatrix} \qquad \text{with all} \quad a_i > 0.$$

Thus, there are 2^N nonzero maximal minors of A_O and

$$\mathcal{M}(A_O) = \{I = \{i_1, \ldots, i_N\} : i_n \in \{2n - 1, 2n\}\}.$$

The N asymptotic line-solitons are identified by the index pairs $\{[2n - 1, 2n] : n = 1, \ldots, N\}$ and the corresponding permutation is $\pi = (2, 1, 4, 3, \ldots, 2N, 2N - 1)$ (or $\pi = (12)(34) \cdots (2N, 2N - 1)$ in the cycle notation). The direction of the n-th soliton is $c_n = \kappa_{2n-1} + \kappa_{2n}$, i.e. $c_1 < c_2 < \ldots < c_N$. Apart from the phase shift of each soliton, the interaction gives rise to a pattern of N intersecting lines in the (x, y)-plane, as shown in Fig. 7.1a.

The letter "O" in the O-type stands for "original" or "ordinary" since the solutions of this type were first found in the Hirota bilinear form in [115] (see also [56]). It is rather surprising that for many years, the soliton solutions of the KP equation are considered to be only of this type. Also note that the vertex operators introduced in [35] generate only this type of KP solitons. An extension of the vertex operator for more general KP solitons is given in [7].

Example 7.2 T-type N-soliton solutions: The simplest example of these solutions can be obtained by choosing the functions in (3.6) as

$$f_n = \partial_x^{n-1} f \qquad \text{with} \quad f(x, y, t) = \sum_{m=1}^{2N} e^{\theta_m(x, y, t)},$$

which yields the coefficient matrix

$$A_T = \begin{pmatrix} 1 & 1 & \cdots & 1 \\ \kappa_1 & \kappa_2 & \cdots & \kappa_{2N} \\ \vdots & \vdots & \ddots & \vdots \\ \kappa_1^{N-1} & \kappa_2^{N-1} & \cdots & \kappa_{2N}^{N-1} \end{pmatrix}.$$

Note that all of the Plücker coordinates are positive, each being equal to a Vandermonde determinant, i.e. $A_T \in Gr(N, M)_{>0}$ so that $\mathcal{M}(A_T) = \binom{[M]}{N}$. These solutions were first investigated in [17] where it was shown that they also satisfy the finite Toda lattice hierarchy. Because of this, we refer to this type of solutions as "T-type" [69]. The n-th asymptotic line-soliton is labeled by the index pair $[n, N + n]$, that is, the corresponding permutation is $\pi = (N + 1, N + 2, \ldots, 2N, 1, 2, \ldots, N)$ (or $\pi = (1, N + 1)(2, N + 2) \cdots (N, 2N))$, whose open chord diagram has the maximum number of crossings $N(N - 1)/2$. It was also shown in [17] that these soliton solutions display phenomena of soliton resonance and web structures as shown in Fig. 7.1b. Notice that all of the asymptotic and intermediate line-solitons interact via three-wave resonances, i.e. all the vertices in the soliton graph are trivalent.

Example 7.3 P-type N-soliton solutions: Yet another type of N-soliton solutions is obtained by

$$f_n(x, y, t) = e^{\theta_n} + a_n e^{\theta_{2N-n+1}}, \quad n = 1, 2, \ldots N,$$

in (3.6). The coefficient matrix is given by

$$A_P = \begin{pmatrix} 1 & 0 & \cdots 0 & 0 & \cdots & 0 & a_1 \\ 0 & 1 & \cdots 0 & 0 & \cdots & a_2 & 0 \\ \vdots & \ddots & \ddots & \vdots & \vdots & \ddots & \ddots & \vdots \\ 0 & \cdots & 0 & 1 & a_N & 0 & \cdots & 0 \end{pmatrix} \quad \text{with} \quad \text{sign}(a_n) = (-1)^{N-n}.$$

Like A_O, there are 2^N nonzero Plücker coordinates and

$$\mathcal{M}(A_P) = \{I \in \{i_1, \ldots, i_N\} : i_n \in \{n, 2N - n + 1\}\}.$$

The n-th asymptotic soliton is identified by the index pair $[n, 2N - n + 1]$ and the corresponding permutation is $\pi = (2N, 2N - 1, \ldots, 2, 1)$ (or $\pi = (1, 2N)(2, 2N - 1) \cdots (N, N + 1))$. The soliton direction is given by $c_n = \kappa_n + \kappa_{2N-n+1}$. Note that the soliton directions are not ordered as in the previous two cases. In fact, taking $c_1 = c_2 = \cdots = c_N = 0$ yields the solutions of the KdV equation [69]. These solutions interact non-resonantly like the O-type N-solitons, i.e., pairwise with an overall phase shift after collision (see Fig. 7.1c).

Since the KP equation was derived for a weak transverse stability problem of the KdV soliton, the solution of the KP equation is considered to be nearly parallel to the *y*-axis. The soliton solution of P-type includes the KdV solitons, and because of this physical setting, the letter "P" stands for "physical" (see [69]).

7.3 Combinatorics of *N*-Soliton Solutions

In the previous section, we have shown that the *N*-soliton solution is characterized by the set of fixed point free involutions of the permutation group S_{2N}. In turn, these involutions of S_{2N} can be enumerated in terms of the various possible arrangements of the pivot and non-pivot columns of the matrix A. In order to describe each involution, we use a chord digram consisting of N chords. Most of the results in this section can be also found in [25, 26].

7.3.1 Linear Chord Diagrams for N-Soliton Solutions

Consider a partition of the integer set $[2N]$ into N distinct 2-element blocks or *pairings*

$$\mathbf{p}_n := [i_n, \ j_n], \quad 1 \le i_n < j_n \le 2N, \quad n = 1, 2, \ldots, N,$$

such that $[2N]$ is a union of the blocks $\mathbf{p}_1, \mathbf{p}_2, \ldots, \mathbf{p}_N$. In combinatorics, such a partition is referred to as a (perfect) *matching* of $[2N]$. We will denote the set of all matchings of $[2N]$ by M_N. The total number of matchings in M_N is given by (see e.g. [18])

$$|\mathsf{M}_N| = 1 \cdot 3 \cdot 5 \ldots (2N - 1) =: (2N - 1)!!.$$

A standard way to represent a matching of M_N, is to mark $2N$ points on a line from left to right labeled by $1, 2, \ldots, 2N$, and join the two points of each pairing $\mathbf{p_n}$ by a chord above the line (see e.g. Fig. 7.2). In terms of the chord diagram in Definition 5.14, it corresponds to a diagram symmetric with respect to the horizontal line.

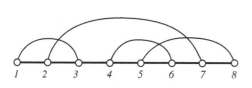
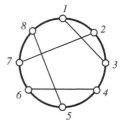

Fig. 7.2 Linear and *circular* chord diagrams for $\pi = (3, 7, 1, 6, 8, 4, 2, 5)$ corresponding to 4-soliton solution

Without loss of generality, the smallest integer i_n from each pairing $\mathbf{p}_n = [i_n, j_n]$ can be arranged in a strictly increasing order $1 = i_1 < i_2 < \ldots i_N \leq 2N - 1$, and the set $I = \{i_1, \ldots, i_N\}$ corresponds to the pivot set of the matrix A. Note that the j_n's are not ordered in general.

Definition 7.1 Let $\mathbf{p}_r$ and $\mathbf{p}_s$ with $r < s$ be distinct pairings (equivalently, a pair of chords). Then,

(a) $\mathbf{p}_r$ and $\mathbf{p}_s$ form an *alignment* or an O-type configuration if $i_r < j_r < i_s < j_s$, i.e. the pairs do not overlap.
(b) $\mathbf{p}_r$ and $\mathbf{p}_s$ form a *crossing* or a T-type configuration if $i_r < i_s < j_r < j_s$, i.e. the pairs partially overlap.
(c) $\mathbf{p}_r$ and $\mathbf{p}_s$ form a *nesting* or a P-type configuration if $i_r < i_s < j_s < j_r$, i.e. the pairs completely overlap.

The names of O-, T-, and P-type of configurations are given as discussed in the previous section. In Fig. 7.2, the pairing $[4, 6]$ forms an alignment (O-type configuration) with $[1, 3]$, a crossing (T-type) with $[5, 8]$, and a nesting (P-type) with the pairing $[2, 7]$. Furthermore, the total number of (pairwise) crossings in Fig. 7.2 is 3 and occur between the pairs $\{[1, 3], [2, 7]\}$, $\{[2, 7], [5, 8]\}$ and $\{[4, 6], [5, 8]\}$. Similarly, there is one nesting, $\{[2, 7], [4, 6]\}$, and two alignments, $\{[1, 3], [4, 6]\}$ and $\{[1, 3], [5, 8]\}$.

It should be clear that the number of alignments, crossings and nestings for any matchings in M_N must add up to the total number of pairwise chord configurations, i.e., $N(N - 1)/2$. One of the earliest results [40, 135] in the enumeration of chord diagrams is that the number of matchings in M_N with *no* crossings is given by the N-th Catalan number $C_N = \frac{1}{N+1}\binom{2N}{N}$, which appears in many combinatorial problems (see, e.g., [120]). Similarly, the number of diagrams in M_N with *no* nestings is also given by C_N. The problem of counting the elements in M_N according to the number of pairwise crossings of chords was considered by Touchard [125], who gave an implicit formula for the generating function in terms of continued fractions. Subsequently, Riordan [106] derived a remarkable explicit formula for the generating function based on Touchard's work. If $cr(X)$ denotes the number of crossings of the element $X \in \mathsf{M}_N$, then the generating function by the number of crossings is defined via the polynomial

$$F_N(q) := \sum_{X \in \mathsf{M}_N} q^{cr(X)}, \qquad 0 \leq cr(X) \leq \tfrac{1}{2}N(N - 1)$$

in the variable q with positive integer coefficients. The Touchard-Riordan formula for $F_N(q)$ is

$$F_N(q) = \frac{1}{(1 - q)^N} \sum_{n=0}^{N} (-1)^n \left[\binom{2N}{N - n} - \binom{2N}{N - n - 1} \right] q^{n(n+1)/2}. \qquad (7.6)$$

The first few polynomials are

$$F_1(q) = 1, \quad F_2(q) = q + 2, \quad F_3(q) = q^3 + 3q^2 + 6q + 5,$$

and it easily follows from (7.6) that the number of non-crossing diagrams is given by

$$F_N(0) = \binom{2N}{N} - \binom{2N}{N-1} = \frac{1}{N+1}\binom{2N}{N},$$

which is the Catalan number C_N mentioned earlier. However, the Touchard-Riordan formula is somewhat mysterious in that it is not obvious from (7.6) that $F_N(q)$ is in fact a polynomial in q of degree $N(N-1)/2$ as implied by its combinatorial origin, or that $F_N(1) = |M_N| = (2N-1)!!$. These assertions follow only after detailed analysis of (7.6) [106] (see also [41]). A purely combinatorial proof of the Touchard-Riordan formula also appeared in Ref. [102] (see also [65]) and its relation to q-Hermite polynomials was investigated in Ref. [59].

7.3.2 Combinatorics of the Chord Diagrams Associated to N-Soliton Solutions

Here we associate an N-soliton equivalence class defined by the set $\mathscr{S} = \{[i_n, j_n]\}_{n=1}^{N}$ of asymptotic line solitons (see Definition 6.5) with the partition $X = \{\mathbf{p}_1, \mathbf{p}_2, \ldots, \mathbf{p}_N\} \in M_N$, where $\mathbf{p}_n := [i_n, j_n]$. Recall from Property 7.1(ii) that the integer set $[2N]$ is a disjoint union of the index sets $I = \{i_1, \ldots, i_N\}$ and $J = \{j_1, \ldots, j_N\}$ with the following ordering among the indices:

(i) $1 = i_1 < i_2 < \ldots < i_N < 2N$,
(ii) $i_n < j_n$ for all $n = 1, 2, \ldots, N$.

An immediate consequence of the above ordering is that

$$n \le i_n \le 2n - 1, \qquad n = 1, \ldots, N, \tag{7.7}$$

since there are at least $n - 1$ indices to the left of i_n, namely $i_1, i_2, \ldots, i_{n-1}$ and at least $2N - 2n + 1$ indices to the right of i_n, namely j_n, i_r, j_r, $r > n$. The N-soliton classification scheme is obtained by considering various statistics over the possible chord configurations for the chord diagrams of M_N. For this purpose, using Definition 7.1, we introduce the following sets, which record the total number of alignments, crossings and nestings for a given chord in any chord diagram of M_N.

Definition 7.2 Let $\mathbf{p}_n = [i_n, j_n]$ be a given chord of a partition $X \in M_N$, and let $B_n := \{\mathbf{p}_r = [i_r, j_r] : i_r < i_n\}$ be the subset of chords originating from the left of $\mathbf{p}_n$ in the linear chord diagram of X.

(a) The set O_n of alignments with the chord $\mathbf{p}_n$ forming O-type configurations and the alignment number $al(X)$ are defined by

$$O_n := \{\mathbf{p}_r = [i_r, j_r] \in B_n : j_r < i_n\}, \qquad al(X) := \sum_{n=1}^{N} |O_n|.$$

(b) The set T_n of crossings with the chord $\mathbf{p}_n$ forming T-type configurations and the crossing number $cr(X)$ are defined by

$$T_n := \{\mathbf{p}_r = [i_r, j_r] \in B_n : i_n < j_r < j_n\}, \qquad cr(X) := \sum_{n=1}^{N} |T_n|.$$

(c) The set P_n of nestings with the chord $\mathbf{p}_n$ forming P-type configurations and the nesting number $ne(X)$ are defined by

$$P_n := \{\mathbf{p}_r = [i_r, j_r] \in B_n : j_r > j_n\}, \qquad ne(X) := \sum_{n=1}^{N} |P_n|.$$

It follows from the above definitions that B_n is the disjoint union of the sets O_n, T_n and P_n, so that $|O_n| + |T_n| + |P_n| = n - 1$ and $al(X) + cr(X) + ne(X) = N(N-1)/2$, which is a count of all possible pairwise chord configurations in the partition X. Note that for O_n, the indices j_r lie in the intervals (i_r, i_{r+1}), $1 \leq r < n$. Hence, $|O_n| = i_n - n$, the number of crossings and nestings with the chord $\mathbf{p}_n$ sum to $|T_n| + |P_n| = (n-1) - (i_n - n) = 2n - i_n - 1$, which depends *only on the pivot index* $i_n \in E$. This observation leads to the following:

Lemma 7.1 *If* $\mathsf{M}(I) \subseteq \mathsf{M}_N$ *denotes the set of all partitions, which have the same (pivot) index set* I, *then the number of partitions of* $\mathsf{M}(I)$ *having* r *crossings and* s *nestings is the coefficient of* $p^s q^r$ *in*

$$m_I(p, q) = \prod_{n=1}^{N} [2n - i_n]_{p,q}, \qquad [n]_{p,q} := \frac{p^n - q^n}{p - q} = \sum_{i+j=n-1} p^i q^j.$$

The degree of both p *and* q *in* $m_E(p, q)$ *is* $N^2 - (i_1 + i_2 + \cdots + i_N)$.

Proof The distribution of crossings and nestings is the sum of $p^{ne(X)} q^{cr(X)}$ over all partitions $X \in \mathsf{M}(I)$. Using Definition 7.2 for $cr(X)$ and $ne(X)$, this distribution can be expressed as

$$\sum_{X \in \mathsf{M}(I)} p^{(|P(1)|+\ldots+|P(N)|)} q^{(|T(1)|+\ldots+|T(N)|)} = \prod_{n=1}^{N} \sum_{l=0}^{2n-i_n-1} p^l q^{2n-i_n-1-l},$$

after interchanging the sum and product, and using the fact that $|T_n| + |P_n| = 2n - i_n - 1$ for $n = 1, \ldots, N$. Since the second sum is precisely $[2n - i_n]_{p,q}$, the formula for $m_I(p, q)$ follows. $\square$

It is easy to verify from the product formula that $m_I(p, q)$ is symmetric in p and q. Consequently, the number of diagrams with r crossings and s nestings is the same as the number of diagrams with s crossings and r nestings [65]. Note also that the enumerating polynomial for the crossings alone is given by $m_I(1, q)$, while

$m_I(p, 1)$ enumerates only the nestings for the chord diagrams of $M(I)$. In order to extend the results of Lemma 7.1 to the entire set M_N, one needs to sum $m_I(p, q)$ over all possible choices of the pivot set I, and the sum is denoted by $\sum_{\{I\}} m_I(p, q)$. Then using Lemma 7.1, the expression for the required generating polynomial is given by

$$F_N(p, q) := \sum_{X \in M_N} p^{ne(X)} q^{cr(X)} = \sum_{\{I\}} m_I(p, q) = \sum_{\substack{1=i_1 < i_2 < \ldots < i_N, \\ n \le i_n \le 2n-1}} \prod_{n=1}^{N} [2n - i_n]_{p,q}.$$

(7.8)

From (7.7) and Lemma 7.1, we know that the degree of p and q in $F_N(p, q)$ is given by

$$ne(X)_{max} = cr(X)_{max} = N^2 - (1 + 2 + \cdots + N) = \frac{N(N-1)}{2}.$$

Furthermore, like $m_I(p, q)$, $F_N(p, q)$ is symmetric in p and q, i.e.,

$$F_N(p, q) = \sum_{r,s=0}^{N(N-1)/2} c_{rs} q^r p^s, \qquad c_{rs} = c_{sr}.$$

The polynomials $F_N(p, q)$ can be determined from a generating function $F(p, q, x)$, which is a formal power series and has the following representation:

Proposition 7.2 *The generating function for $F_N(p, q)$ is the Stieltjes-type continued fraction, i.e.*

$$F(p, q, x) := \sum_{N=0}^{\infty} F_N(p, q) x^N = \cfrac{1}{1 - \cfrac{x\,[1]_{p,q}}{1 - \cfrac{x\,[2]_{p,q}}{1 - \cfrac{x\,[3]_{p,q}}{1 - \cdots}}}} \qquad \text{with} \quad F_0(p, q) := 1,$$

where $[n]_{p,q}$ is defined in Lemma 7.1.

Proof First consider the set $I := \{1 = i_1 < \cdots < i_N : i_k \le 2k - 1, k = 1, \ldots, N\}$. Note that I can be decomposed into distinct subsets when $i_n = 2n - 1$. One has

$$I = \bigcup_{n=0}^{N-1} \left(I_n \cup \hat{I}_n \right),$$

where $I_n := \{1 = i_1 < \cdots < i_n : i_k \le 2k - 1, k = 1, \ldots, n\}$ for $n \ne 0$ can be viewed as the n-truncates of the original set $I = I_N$, $I_0 = \emptyset$ and

$$\hat{I}_n := \{2n+1 = i_{n+1} < \cdots < i_N : 2n+k \le i_{n+k} < 2(n+k)-1, k = 1, \ldots, N-n\}.$$

The set $\hat{I}_n$ can be re-expressed as

$$I'_{N-n} = \{1 = i'_1 < \cdots < i'_{N-n} : k \leq i'_k < 2k - 1, k = 1, \ldots, N - n\}$$

by shifting and relabeling the indices as $i_{n+k} := i'_k + 2n$. Note however that I'_{N-n} (with all $i'_k < 2k - 1$) is *not* the same as I_{N-n} (with all $i_k \leq 2k - 1$).

From (7.8), we have

$$F_N(p, q) = \sum_{\{I\}} \prod_{k=1}^{N} [2k - i_k]_{p,q} = \sum_{n=0}^{N-1} F_n(p, q) C_{N-n}(p, q), \qquad (7.9)$$

where $C_n(p, q) = \sum_{\{I'_n\}} \prod_{k=1}^{n} [2k - i'_k]_{p,q}$. We then introduce the power series

$$F(p, q, x) = \sum_{N=0}^{\infty} F_N(p, q) x^N \quad \text{with} \quad F_0(p, q) := 1,$$

$$C(p, q, x) = \sum_{N=1}^{\infty} C_N(p, q) x^N.$$

Using (7.9) in the power series, one finds that $F(p, q, x) - 1$ equals the product $F(p, q, x) C(p, q, x)$, which implies

$$F(p, q, x) = \frac{1}{1 - C(p, q, x)}. \qquad (7.10)$$

Next, define the associated polynomials

$$F_n(p, q; l) := \sum_{\{I_n\}} \prod_{k=1}^{n} [2k - i_k + l]_{p,q}, \quad C_n(p, q; l) := \sum_{\{I'_n\}} \prod_{k=1}^{n} [2k - e'_k + l]_{p,q},$$

so that $F_n(p, q; 0) = F_n(p, q)$ and $C_n(p, q; 0) = C_n(p, q)$. The corresponding power series $F(p, q; l, x)$ and $C(p, q; l, x)$ are defined similarly to $F(p, q, x)$ and $C(p, q, x)$ above, and they also satisfy Eq. (7.10). Furthermore, for $n > 1$ the associated polynomials satisfy the relation

$$C_n(p, q; l) = \sum_{\{E_n\}} \prod_{k=1}^{n} [2k - i'_k + l]_{p,q} = [l + 1]_{p,q} \sum_{\{E_n\}} \prod_{k=2}^{n} [2k - i'_k + l]_{p,q}$$

$$= [l + 1]_{p,q} \sum_{\{E_{n-1}\}} \prod_{j=1}^{n-1} [2j - i_j + (l + 1)]_{p,q} = [l + 1]_{p,q} F_{n-1}(p, q; l + 1),$$

$$\qquad (7.11)$$

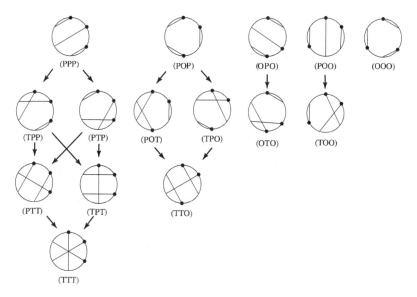

Fig. 7.3 The *circular* chord diagrams for 3-soliton solutions. The *dots* indicate the pivots (i_1, i_2, i_3), and the ordered letters below each diagram indicate the type of interactions in $(\mathbf{p}_1\mathbf{p}_2, \mathbf{p}_2\mathbf{p}_3, \mathbf{p}_3\mathbf{p}_1)$ with the soliton pairing $\mathbf{p}_n = [i_n, j_n]$. The total number of the diagrams having the same number of crossings is given by the generating function $F_3(q) = q^3 + 3q^2 + 6q + 5$

after an appropriate index shift, $k = j + 1$ and relabeling $i'_{j+1} = i_j + 1$ so that $j \le i_j \le 2j - 1$ for $j = 1, \ldots n - 1$. As a result, the set I'_n is changed to the set I_{n-1}. The formal power series constructed from the first and last expressions in Eq. (7.11) satisfies $C(p, q; l, x) = x[l + 1]_{p,q} F(p, q; l + 1, x)$. From the analogue of (7.10) for the associated functions $F(p, q; l, x)$ and $C(p, q; l, x)$, one therefore obtains

$$F(p, q; l, x) = \frac{1}{1 - [l + 1]_{p,q} x F(p, q; l + 1, x)}.$$

This yields the continued fraction representation for $F(p, q, x) = F(p, q; 0, x)$. $\square$

We can graphically illustrate the results of Proposition 7.2 for $N = 3$ in terms of the corresponding chord diagrams. There are 15 chord diagrams, which are displayed in Fig. 7.3. They are characterized by

$$F_3(p, q) = (1 + 2p + p^2 + p^3) + (2 + 2p + 2p^2)q + (1 + 2p)q^2 + q^3.$$

Remark 7.2 It is intriguing to note that the generating function $F(0, q, x)$ for the non-nesting chord diagrams is related to the moment generating function for a certain class of q-orthogonal polynomials studied in Ref. [4] (see also [58]). A particularly interesting consequence of this relation is that the function $F(0, q, -q)$ has a Rogers-Ramanujan interpretation. Let ϕ_1 and ϕ_2 be certain modular forms of weight $\frac{1}{5}$ for the level-5 principal modular group $\Gamma(5) < PSL(2, \mathbb{Z})$; namely,

$$\phi_1(q) = \frac{1}{\eta(q)^{3/5}} \sum_{n \in \mathbb{Z}} (-1)^n q^{(10n+1)^2/40}, \quad \phi_2(q) = \frac{1}{\eta(q)^{3/5}} \sum_{n \in \mathbb{Z}} (-1)^n q^{(10n+3)^2/40},$$

(7.12)

where $\eta(q) = q^{1/24} \prod_{n=1}^{\infty} (1 - q^n)$ is the Dedekind η-function. It is well-known that the modular forms ϕ_1 and ϕ_2 admit infinite product representations, which constitute the Rogers-Ramanujan identities. Accordingly, $F(0, q, -q)$ can be represented as the following quotient:

$$F(0, q, -q) = q^{-1/5} \frac{\phi_2}{\phi_1} = \prod_{n=0}^{\infty} \frac{(1 - q^{5n+1})(1 - q^{5n+4})}{(1 - q^{5n+2})(1 - q^{5n+3})} = \cfrac{1}{1 + \cfrac{q}{1 + \cfrac{q^2}{1 + \cfrac{q^3}{1 + \cdots}}}}.$$

This is the Rogers-Ramanujan continued fraction [104, 107].

Problems

7.1 Give an example of the matrix A in $\mathrm{Gr}(2, 5)_{\geq 0}$ and its dual matrix B in $\mathrm{Gr}(3, 5)_{\geq 0}$ such that all the Plücker coordinates are equal to each other, i.e. $\Delta_I(A) = \Delta_J(B)$ for all $I \cup J = [5]$.

7.2 Using the duality between $\mathrm{Gr}(N, M)_{\geq 0}$ and $\mathrm{Gr}(M - N, M)_{\geq 0}$, prove Theorem 6.1 for $y \ll 0$.

7.3 Find the $\mathcal{J}$-diagrams (equivalently the pairs (v, w)) corresponding to the KP solitons in Fig. 7.1.

7.4 The *orthogonal* Grassmannian, denoted by $\mathrm{OGr}(N, 2N)$, may be defined by

$$\mathrm{OGr}(N, 2N) = \left\{ A \in \mathrm{Gr}(N, 2N) : A \eta A^T = 0 \right\},$$

where $\eta = \mathrm{diag}(1, -1, 1, -1, \ldots, 1, -1)$ (see [68]).
   Show that the orthogonality condition $A \eta A^T = 0$ implies that

(a) $\Delta_I(A) = \Delta_J(A)$ for $I \cup J = [2N]$, and
(b)   the dimension of $\mathrm{OGr}(N, 2N)$ is $N(N - 1)/2$.

Note here that item (a) implies that the KP soliton generated from $\mathrm{OGr}(N, 2N)$ is nothing but the N-soliton solution introduced in Sect. 7.2.

Also show that $OGr(2, 4)_{\geq 0}$ can be expressed as

$$OGr(2, 4)_{\geq 0} = \left\{ A = \begin{pmatrix} 1 & \tanh\phi & 0 & -\operatorname{sech}\phi \\ 0 & \operatorname{sech}\phi & 1 & \tanh\phi \end{pmatrix} : 0 \leq \phi \leq \infty \right\}.$$

Note that at $\phi = 0$, the element A gives a P-type matrix, and at $\phi = \infty$, A gives an O-type matrix (see Sect. 7.2).

Find an explicit parametrization for $OGr(3, 6)_{\geq 0}$. That is, for each chord diagram in Fig. 7.3, give the corresponding 3×6 matrix. What are the labeled J-diagrams corresponding to these chord diagrams?

Chapter 8
Soliton Graphs

Abstract In this chapter, we explicitly construct the soliton graph $\mathscr{C}_t(\mathscr{M}(A))$ for an irreducible matrix A from $\mathscr{P}_{v,w}^{>0} \subset \mathrm{Gr}(N, M)_{\geq 0}$. In particular, we provide an algorithm that constructs the soliton graphs for the matrix A and give coordinates for all of the trivalent vertices, which then allows one to completely describe the soliton graph. Most of this chapter will be devoted to the case when $t < 0$, with the final section explaining how the same ideas can be applied to the case when $t > 0$.

8.1 Soliton Graphs $\mathscr{C}_\pm(\mathscr{M}(A))$

Let us first recall the piecewise linear function $f_{\mathscr{M}(A)}(x)$ defined in (6.2),

$$f_{\mathscr{M}(A)}(x, y, t) = \max\{\Theta_I(x, y, t) : I \in \mathscr{M}(A)\} \quad \text{with} \quad \Theta_I(x, y, t) = p_I x + q_I y + \omega_I(t),$$

where $(p_I, q_I, \omega_I(t))$ for $I \in \mathscr{M}(A)$ are given by

$$p_I = \sum_{i \in I} \kappa_i, \qquad q_I = \sum_{i \in I} \kappa_i^2, \qquad \omega_I(t) = \sum_{i \in I} \kappa_i^3 t. \tag{8.1}$$

Notice that for each t, the set $\{\Theta_I(x, y, t) : I \in \mathscr{M}(A)\}$ gives an arrangement of the planes $z = \Theta_I(x, y, t)$ over the xy-plane and $f_{\mathscr{M}(A)}(x, y, t)$ gives the dominant plane at each point $(x, y) \in \mathbb{R}^2$.

Remark 8.1 As we explained in Chap. 3, one can also discuss soliton solutions for different equations, such as 2-dimensional Toda lattice and the Davey-Stewartson (DS) equation. To obtain the soliton solutions for each of these equations, we take the set of the parameters $\{(p_i, q_i) : i = 1, \ldots, M\}$ from a particular conic curve. For example, the DS soliton can be obtained by choosing $(p_i, q_i, \omega_i(t)) = (-\sin \psi_i, \cos \psi_i, -\sin(2\psi_i)t)$. In this chapter, we discuss only the KP solitons (other cases may be a good project for a graduate student).

© The Author(s) 2017
Y. Kodama, *KP Solitons and the Grassmannians*,
SpringerBriefs in Mathematical Physics 22, DOI 10.1007/978-981-10-4094-8_8

The soliton graph $\mathscr{C}_t(\mathscr{M}(A))$ for fixed time t was defined in Definition 6.1, i.e.

$$\mathscr{C}_t(\mathscr{M}(A)) := \{\text{the locus in } \mathbb{R}^2 \text{ where } f_{\mathscr{M}(A)}(x, y, t) \text{ is not linear}\}.$$

The main goal of this chapter is to construct the soliton graphs for large $|t|$. As mentioned in Remark 6.1, there is only one soliton graph for each $t > 0$ or $t < 0$, which we denote as $\mathscr{C}_+(\mathscr{M}(A))$ or $\mathscr{C}_-(\mathscr{M}(A))$. A key step for this purpose is to determine a region in the xy-plane where one of the terms $\Theta_I(x, y, t)$ dominates over others.

In the soliton graph, one line-soliton of $[i, j]$-type is expressed by a line,

$$L_{[i,j]}(t): \quad x + (\kappa_i + \kappa_j)y + (\kappa_i^2 + \kappa_j^2 + \kappa_i\kappa_j)t = 0.$$

This is given by the balance between two dominant phases Θ_I and Θ_J with $J = (I \setminus \{i\}) \cup \{j\} \in \mathscr{M}(A)$, i.e. $\theta_i = \theta_j$ with $\theta_i = \kappa_i x + \kappa_i^2 y + \kappa_i^3 t$.

A Y-soliton of resonant solution is given by the balance of three dominant phases. Let $\mathbf{v}_{[i,j,k]}$ denote the interaction point associated with line-solitons of $[i, j]$-, $[j, k]$- and $[i, k]$-types with $i < j < k$. There are two types of resonant interactions as shown in the figure below (see also Remark 1.4, and Figs. 1.3 and 1.5):

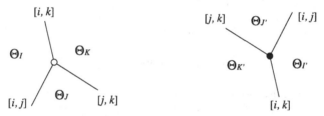

In the figure above, each resonant interaction point is marked by either a *white* vertex or a *black* vertex depending on the index set of dominant phases which are characterized as follows:

(a) For the white vertex, denoted by $\mathbf{v}_{[i,j,k]}^\circ$ with $i < j < k$, the index set $\{I, J, K\} \subset \mathscr{M}(A)$ is given by

$$I = L_0 \cup \{i\}, \qquad J = L_0 \cup \{j\}, \qquad K = L_0 \cup \{k\},$$

for some common $(N - 1)$-index set L_0.

(b) For the black vertex, denoted by $\mathbf{v}_{[i,j,k]}^\bullet$ with $i < j < k$, the index set $\{I', J', K'\} \subset \mathscr{M}(A)$ is given by

$$I' = K_0 \setminus \{i\}, \qquad J' = K_0 \setminus \{j\}, \qquad K' = K_0 \setminus \{k\},$$

for some common $(N + 1)$-index set K_0.

The coordinates $(x_{[i,j,k]}, y_{[i,j,k]})$ of the intersection point for both types can be calculated from the balance $\theta_i = \theta_j = \theta_k$, i.e.

$$x_{[i,j,k]} = (\kappa_i\kappa_j + \kappa_i\kappa_k + \kappa_j\kappa_k)\,t, \quad y_{[i,j,k]} = -(\kappa_i + \kappa_j + \kappa_k)\,t. \qquad (8.2)$$

Definition 8.1 A resonant interaction point $\mathbf{v}_{[a,b,c]}$ is said to be *visible* if we have dominant relations at $\mathbf{v}_{[a,b,c]}$

$$\Theta_{I_a} = \Theta_{I_b} = \Theta_{I_c} > \Theta_J \quad \text{for any} \quad J \in \mathcal{M}(A) \setminus \{I_a, I_b, I_c\},$$

where the index set $\{I_a, I_b, I_c\}$ is either the case (i) above, i.e. $I_a = L_0 \cup \{a\}$ etc., or the case (ii), i.e. $I_a = K_0 \setminus \{a\}$ etc.

Let z_i denote the value of $\theta_i(x, y, t) = \kappa_i x + \kappa_i^2 y + \kappa_i^3 t$ at the vertex $\mathbf{v}_{[a,b,c]}$,

$$z_i = \theta_i(x_{[a,b,c]}, y_{[a,b,c]}, t) \quad \text{for} \quad i \in [M].$$

that is, z_i gives the height of the plane $z = \theta_i(x, y, t)$ at the vertex $\mathbf{v}_{[a,b,c]}$. In particular, let $z_{[a,b,c]}$ denote the value $z_a = z_b = z_c$, which is given by $z_{[a,b,c]} = \kappa_a\kappa_b\kappa_c\,t$. Then the following lemma is easy to verify.

Lemma 8.1 *At the intersection point $\mathbf{v}_{[a,b,c]}$, we have*

$$z_i - z_{[a,b,c]} = t\,(\kappa_i - \kappa_a)(\kappa_i - \kappa_b)(\kappa_i - \kappa_c).$$

This lemma can be used to check whether the point $\mathbf{v}_{[a,b,c]}$ is visible or not.

Example 8.1 Consider the case with $M = 4$ and $N = 1$ (see Fig. 1.4). With the order $\kappa_1 < \cdots < \kappa_4$, Lemma 8.1 implies $z_4 - z_{[1,2,3]} > 0$ for $t > 0$, that is, the plane $P_4 := \{z = \theta_4(x, y, t)\}$ is above the point $\mathbf{v}_{[1,2,3]}$, hence $\mathbf{v}_{[1,2,3]}$ is not visible. We also have $z_1 - z_{[2,3,4]} < 0$ and $z_3 - z_{[1,2,4]} < 0$ for $t > 0$, which implies $\mathbf{v}_{[2,3,4]}$ and $\mathbf{v}_{[1,2,4]}$ are visible. See the right figure in Fig. 8.1.

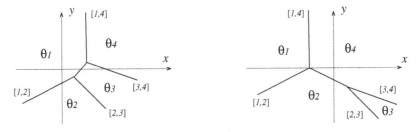

Fig. 8.1 Soliton graphs for $\text{Gr}(1, 4)_{\geq 0}$. The *left* figure shows the graph for $t < 0$ and the *right* one for $t > 0$. The κ-parameters are $(-2, 0, 1, 2)$. Each θ_i shows the dominant plane $z = \theta_i(x, y, t)$ in that region (see Fig. 3.5)

Now we have the following proposition:

Proposition 8.1 *For $t < 0$, any trivalent vertex in the reduced pipedream is visible. That is, there exists a set of indices $\{J_a, J_b, J_c\}$ such that the corresponding phases are the dominant phases and satisfy $\Theta_{J_a} = \Theta_{J_b} = \Theta_{J_c}$ at $\mathbf{v}_{[a,b,c]}$. If the vertex is white, then the indices are expressed by $J_a = L_0 \cup \{a\}$, $J_b = L_0 \cup \{b\}$ and $J_c = L_0 \cup \{c\}$ with some $(N-1)$ common index set L_0. If the vertex is black, then the indices are expressed by $J_a = K_0 \setminus \{a\}$, $J_b = K_0 \setminus \{b\}$ and $J_c = K_0 \setminus \{c\}$ with some $(N+1)$ common index set K_0.*

Proof Here, we consider only the case with a white vertex (the case with a black vertex can be treated in a similar way).

In the figure below, we illustrate a J-diagram and pick a white vertex in a box, referred to as "box ac", where a is a pivot index and c is a non-pivot index. Let $\{I_a, I_b, I_c\}$ be the set of indices around the vertex, which can be calculated from the pipedream (see Sect. 5.3.2). In particular, the index I_c can be calculated from

$$I_c = v_{ac}^{in}(w_{ac}^{in})^{-1}\{i_1, \ldots, i_N\},$$

where the pair of permutations $(v_{ac}^{in}, w_{ac}^{in})$ is defined in Definition 5.12. That is, in the J-diagram, place white stones in all the boxes of the rows above box ac and the columns left of box ac. The new J-diagram represents the elements in $\mathscr{P}_{v_{ac}^{in}, w_{ac}^{in}}^{>0}$. Then the other indices are given by

$$I_a = L_0 \cup \{a\}, \quad I_b = L_0 \cup \{b\} \qquad \text{with} \quad L_0 = I_c \setminus \{c\}.$$

The index set L_0 can be written in the form

$$L_0 = \{i_1 < \cdots < i_k < p_1 < \cdots < p_l < i_{k+l+2} < \cdots < i_N\},$$

where i_n's are the pivot indices and we set $a = i_{k+1} < p_1 < \cdots < p_l < c < i_{k+l+2}$ and $i_{k+l+1} < c$. Recall that I_a is the lexicographically maximal element of the index set of the form

$$\{i_1 < \cdots < i_k < a = i_{k+1} < q_1 < \cdots < q_l < i_{k+l+2} < \cdots < i_N\} \in \mathscr{M}(A). \quad (8.3)$$

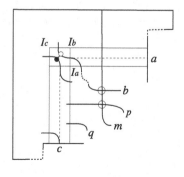

$$a < b < m < c$$
$$a < p < m < q < c$$

$$a, p, q \in I_a, \quad b, m \notin I_a$$
$$b, p, q \in I_b, \quad a, m \notin I_b$$
$$c, p, q \in I_c, \quad a, b, m \notin I_c$$

The white vertex in the box ac corresponds to a resonant interaction point $\mathbf{v}^\circ_{[a,b,c]}$ in the xy-plane. To show the visibility of $\mathbf{v}_{[a,b,c]}$, we look for the dominant phases $\Theta_{J_a} = \Theta_{J_b} = \Theta_{J_c}$, where $J_a = L'_0 \cup \{a\}$, $J_b = L'_0 \cup \{b\}$ and $J_c = L'_0 \cup \{c\}$ for some L'_0. Note in general that $L'_0 \neq L_0$.

Let us first consider the case with $v = 1$ (i.e. the matrix A is an element of a top Schubert cell). In this case, there is no such index shown as m in the figure above. Then the indices $\{p_1 < \cdots < p_l\}$ in L_0 are given by $p_k = b + k$, and $c = p_l + 1 = b + l + 1$. From Lemma 8.1, we have $z_i - z_{[a,b,c]} < 0$ for all $a < i < b$. This implies that $\Theta_{I_a} = \Theta_{I_b} = \Theta_{I_c}$ gives the dominant phase balance at the resonant point $\mathbf{v}^\circ_{[a,b,c]}$.

Now we consider the case having an index m such that $m \notin L_0$ and $a < m < c$. If $\theta_m > \theta_p$ for some $p \in I_a$ at $\mathbf{v}^\circ_{[a,b,c]}$, we have $\Theta_{(I_a \setminus \{p\}) \cup \{m\}} > \Theta_{I_a}$. Then one needs to find the dominant phases with $\{J_a, J_b, J_c\}$ for the visibility of the point $\mathbf{v}^\circ_{[a,b,c]}$. We have the following two cases:

(a) For $p \in L_0$, assume $a < p < m$ (see the figure above). Then we have $(I_a \setminus \{p\}) \cup \{m\} > I_a$ (lexicographical order), and this means that

$$\Delta_{(I_a \setminus \{p\}) \cup \{m\}} = 0,$$

   (see Remark 5.5). So there is no phase Θ_J with $J = (I_a \setminus \{p\}) \cup \{m\}$ (i.e. $J \notin \mathcal{M}(A)$).

(b) For $q \in L_0$, assume that $m < q < c$ and $\theta_m > \theta_q$ at the point $\mathbf{v}^\circ_{[a,b,c]}$. If $\Delta_{J_a} \neq 0$ with $J_a := (I_a \setminus \{q\}) \cup \{m\}$, then $\Theta_{J_a} > \Theta_{I_a}$ at $\mathbf{v}^\circ_{[a,b,c]}$. In this case, we can show that we also have

$$\Delta_{J_b} \neq 0 \quad \text{and} \quad \Delta_{J_c} \neq 0, \tag{8.4}$$

   where $J_b := (I_b \setminus \{q\}) \cup \{m\}$ and $J_c := (I_c \setminus \{q\}) \cup \{m\}$, and these give the dominant phases,

$$\Theta_{J_a} = \Theta_{J_b} = \Theta_{J_c} \quad \text{at} \quad \mathbf{v}_{[a,b,c]}.$$

To show (8.4), we use the Plücker relation,

$$\Delta_{J_a} \Delta_{I_c} - \Delta_{I_a} \Delta_{J_c} + \Delta_{Q_0 \cup \{a,c\}} \Delta_{Q_0 \cup \{m,q\}} = 0,$$

where $Q_0 := L_0 \setminus \{q\}$. Since $Q_0 \cup \{m, q\} = (I_a \setminus \{a\}) \cup \{m\} > I_a$ in the set (8.3), we have $\Delta_{Q_0 \cup \{m,q\}} = 0$. Then we have $\Delta_{J_a} \Delta_{I_c} = \Delta_{I_a} \Delta_{J_c}$. Similarly, one can show that $\Delta_{J_a} \Delta_{I_b} = \Delta_{I_a} \Delta_{J_b}$. That is, we have

$$\Delta_{J_b} = \frac{\Delta_{I_b}}{\Delta_{I_a}} \Delta_{J_a} \quad \text{and} \quad \Delta_{J_c} = \frac{\Delta_{I_c}}{\Delta_{I_a}} \Delta_{J_a}.$$

This proves the proposition. $\square$

8.2 Construction of the Soliton Graphs $\mathscr{C}_{\pm}(\mathscr{M}(A))$

For given matrix $A \in \mathscr{P}^{>0}_{v,w} \subset \mathrm{Gr}(N, M)_{\geq 0}$, we propose an algorithm to construct the soliton graphs $\mathscr{C}_t(\mathscr{M}(A))$. We first consider the case for $t < 0$.

8.2.1 The Soliton Graph $\mathscr{C}_-(\mathscr{M}(A))$

First note that for $t \ll 0$, we have the dominant relation for any fixed (finite) point in $(x, y) \in \mathbb{R}^2$,

$$\theta_1(x, y, t) \gg \theta_2(x, y, t) \gg \cdots \gg \theta_M(x, y, t),$$

and the phase Θ_I with the pivot set I (i.e. the lexicographically minimal element in $\mathscr{M}(A)$) is dominant for any values of y and for sufficiently large negative values of x.

We then propose the following algorithm to construct the soliton graph $\mathscr{C}_-(\mathscr{M}(A))$ associated to $A \in \mathscr{P}^{>0}_{v,w}$ based on the $\mathcal{J}$-diagram.

Algorithm 8.1 From the $\mathcal{J}$-diagram to the soliton graph $\mathscr{C}_-(\mathscr{M}(A))$:

1. Start with the $\mathcal{J}$-diagram which can be constructed by the pair (v, w) or the permutation $\pi = vw^{-1}$.
2. Draw the reduced pipedream. Note that the numbers of trivalent vertices are given by $d - N$ for the white ones and $d - (M - N)$ for the black ones with $d = \ell(w) - \ell(v)$, the dimension of $\mathscr{P}^{>0}_{v,w}$.
3. Give the coordinates for all trivalent vertices in the form $\mathbf{v}_{[a,b,c]} = (x_{[a,b,c]}, y_{[a,b,c]})$ for $a < b < c$ (see (8.2)),

$$x_{[a,b,c]} = (\kappa_a \kappa_b + \kappa_a \kappa_c + \kappa_b \kappa_c)t, \qquad y_{[a,b,c]} = -(\kappa_a + \kappa_b + \kappa_c)t.$$

4. Plot these vertices on the xy-plane and draw three lines incident to each vertex. If the vertex is white, then draw the line $L_{[a,c]}$ for $y > y_{[a,b,c]}$ and two lines $L_{[a,b]}$ and $L_{[b,c]}$ for $y < y_{[a,b,c]}$. If the vertex is black, then draw $L_{[a,b]}, L_{[b,c]}$ for $y > y_{[a,b,c]}$ and $L_{[a,c]}$ for $y < y_{[a,b,c]}$. See the figures above (8.2). Here the line $L_{[i,j]}$ corresponding to the line-soliton of $[i, j]$-type is given by

$$L_{[i,j]} : \; x + (\kappa_i + \kappa_j)y + (\kappa_i^2 + \kappa_j^2 + \kappa_i \kappa_j)t = 0.$$

5. If there is a path with a pair of indices $[i, j]$ in the pipedream which does not connect to any vertex, then draw the infinite line $L_{[i,j]}$ (i.e. $L_{[i,j]}$ corresponds to a line-soliton without any resonant interaction with other solitons).

Then we have the main theorem in this chapter:

Theorem 8.2 *Let A be an $N \times M$ irreducible matrix from a point of $\mathscr{P}_{v,w}^{>0} \subset$ Gr$(N, M)_{\geq 0}$. Then the graph obtained using the Algorithm 8.1 for A is the soliton graph $\mathscr{C}_-(\mathscr{M}(A))$.*

Proof Let L be the J-diagram corresponding to the matrix A. We assume A to be in a row echelon form with the pivot set $\{i_1 < i_2 < \cdots , < i_N\}$. We then consider a series of J-diagrams $(L^{(1)}, L^{(2)}, \ldots, L^{(N)})$ with $L^{(N)} = L$ where $L^{(k)}$ is given by placing white stones in the first $N - k$ rows of the J-diagram L. The main strategy is to use Theorem 6.1 and an induction to construct the corresponding soliton graphs $\mathscr{C}^{(k)} := \mathscr{C}_-(\mathscr{M}(A^{(k)}))$, where $A^{(k)}$ is given by setting all zeros except the pivots for the first $N - k$ rows of the matrix A. That is, each $L^{(k)}$ is the J-diagram corresponding to the matrix $A^{(k)}$ and the matroid $\mathscr{M}(A^{(k)})$ is expressed by

$$\mathscr{M}(A^{(k)}) = \mathscr{M}(A) \cap \left(\bigcup_{\{p_1,\ldots,p_k\}} \{i_1 < \cdots < i_{N-k} < p_1 < \cdots < p_k\} \right),$$

where the sum is over all index sets of $\{p_1, \ldots, p_k\} \subset \{i_{N-k+1}, i_{N-k+1} + 1, \ldots, M\}$. Note that the pivot set $I = \{i_1, \cdots, i_N\}$ is the lexicographically minimal element for every $A^{(k)}$, and we denote $\mathscr{M}(A^{(0)}) = \{I\}$. Note also that for any values of y, the phase $\Theta_I(x, y, t)$ becomes dominant for sufficiently large negative values of x. This implies that for each $\mathscr{C}^{(k)}_-$, there is a polyhedral subset $\mathscr{R}^{(k)} \subset \mathbb{R}^2$ including a region where Θ_I is dominant and the boundary of $\mathscr{R}^{(k)}$ is formed by line-solitons of $[i_{N-k+1}, q_l]$-types. More precisely, for each t, the polyhedral subset $\mathscr{R}^{(k)}$ is given by

$$\mathscr{R}^{(k)} = \{ \text{ the region where } \Theta_J(x, y, t) \text{ is dominant for } J \in \mathscr{M}(A^{(k-1)})\}.$$

For example, at $k = 1$, $\mathscr{M}(A^{(1)}) = \mathscr{M}(A) \cap \{\{i_1, \ldots, i_{N-1}, p_1\} : i_N \leq p_1 \leq M\}$, and $A^{(1)}$ can be considered as an element of Gr$(1, N^{(1)})$ where $N^{(1)}$ is the number of nonzero columns larger than or equal to i_N in $A^{(1)}$. The polyhedral subset $\mathscr{R}^{(1)}$ is the region where Θ_I with the pivot set I is dominant and the boundary is given by line-solitons with $[i_N, p]$ with some $i_N < p < M$. From Theorem 6.1, the asymptotic line-solitons for $A^{(1)}$ are as follows:

(a) For $y \gg 0$, there is one soliton of $[i_N, p_l]$-type for some l.
(b) For $y \ll 0$, there are l solitons of $[i_N, p_1]$-, $[p_1, p_2]$-, $\ldots$, $[p_{l-1}, p_l]$-types.

Note here that for irreducible matrix A, $p_k = i_N + k$ and the entries of A satisfy $a_{N,m} = 0$ for $m > N + l$, i.e. $N^{(1)} = l + 1$. Each soliton of $[p_k, p_{k+1}]$-type in $y \ll 0$ has a resonant interaction with solitons of $[i_N, p_k]$- and $[i_N, p_{k+1}]$-types and the interaction point is given by $\mathbf{v}_{[i_N, p_k, p_{k+1}]}$, which is identified as a white vertex in the $L^{(1)}$ diagram. From Proposition 8.1, those trivalent vertices are visible and they give the information on the dominant phases for the regions in the complement of $\mathscr{C}^{(1)}_-$.

Now consider the case of $A^{(2)}$. Again from Theorem 6.1, one can identify the asymptotic solitons. Some of them have resonant interaction with $[p_k, p_{k+1}]$-solitons constructed in the previous step. These resonant points can be identified with the trivalent vertices in the $(N-1)$-th row in the $\mathcal{J}$-diagram. Notice that the polyhedral region $\mathscr{R}^{(2)}$ consists of the dominant phases observed in the previous step and the boundary of $\mathscr{R}^{(2)}$ are given by solitons of $[i_{N-1}, q]$-types with some $i_{N-1} < q < M$. In particular, the resonant points at the boundary of $\mathscr{R}^{(2)}$ have the form $\mathbf{v}_{[i_{N-1}, p, q]}$ for some $i_{N-1} < p < q < M$, and the interaction points $\mathbf{v}_{i_N, i, j}$ in the previous step are all in $\mathscr{R}^{(1)}$.

Continuing this inductive process, one can obtain the soliton graph $\mathscr{C}_-(\mathscr{M}(A))$. Notice in this process that we do not see any new resonant interactions other than those identified as the trivalent vertices in the pipedream (see Fig. 8.2). This completes the proof. $\qquad\square$

The matrix A in Fig. 8.2 is from a point in $\mathscr{P}^{>0}_{v,w} \subset \mathrm{Gr}(4, 8)_{\geq 0}$ where (v, w) is given by

$$w = s_6 s_7 s_4 s_5 s_6 s_2 s_3 s_4 s_5 s_1 s_2 s_3 s_4, \qquad v = s_6 s_4 s_1 s_3 s_4.$$

The set of trivalent vertices in the pipedream is given by

$$\{\mathbf{v}^{\circ}_{[6,7,8]}, \mathbf{v}^{\circ}_{[4,5,7]}, \mathbf{v}^{\circ}_{[2,3,5]}, \mathbf{v}^{\circ}_{[2,4,8]}, \mathbf{v}^{\bullet}_{[4,6,7]}, \mathbf{v}^{\bullet}_{[2,7,8]}, \mathbf{v}^{\bullet}_{[2,3,5]}, \mathbf{v}^{\bullet}_{[1,3,5]}\}.$$

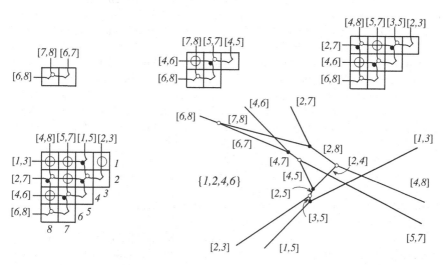

Fig. 8.2 Induction to generate the soliton graph $\mathscr{C}_-(\mathscr{M}(A))$ for $A \in \mathrm{Gr}(4, 8)_{\geq 0}$. Notice that each colored vertex in the Le-diagram corresponds to a unique interaction point in the soliton graph and each pipe in the pipe dream gives a unique line-soliton in the graph. Identify the polyhedral regions $\mathscr{R}^{(k)}$ for $k = 1, 2, 3, 4$

Following Algorithm 8.1, we can obtain the soliton graph $\mathscr{C}_-(\mathscr{M}(A))$. Here we use the κ-parameters $(\kappa_1, \ldots, \kappa_8) = (-1.5, -1, -0.5, 0, 0.5, 1, 1.5, 2.5)$. The detail of the construction of the graph is left for the readers.

8.2.2 The Soliton Graph $\mathscr{C}_+(\mathscr{M}(A))$

We here discuss the case for $t > 0$ by using the example in Fig. 8.2. First we note that for $t > 0$, the phase Θ_J with the lexicographically maximal element $J \in \mathscr{M}(A)$ is dominant at any finite fixed point $(x, y) \in \mathbb{R}^2$. Then we consider a *time-dual* $\mathcal{J}$-diagram, denoted by $L^+(A)$, which is defined as follows: First we draw the permutation $\pi = vw^{-1}$ in two ways as illustrated in the figure below.

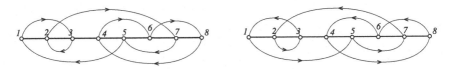

Each arrow in the left panel expresses $i \to \pi(i)$ and in the right panel expresses $i \to \pi^{-1}(i)$. Recall that each pivot index i_k of A is defined by $i_k < \pi(i_k)$ for $k = 1, \ldots, N$, and that i_k gives the minimal index of the nonzero entry of the k-th row of the matrix A in the row echelon form, i.e. $a_{i_k,l} = 0$ if $l < i_k$ and $a_{i_k,i_k} \neq 0$. We call the pivot i_k the *left* pivot. From the right panel, we define the *right* pivot i_k' if $i_k' > \pi^{-1}(i_k')$. In this example, the set of right pivots is $\{8, 7, 6, 3\}$. Then the time-dual $\mathcal{J}$-diagram $L^+(A)$ can be defined as in the left panel in Fig. 8.3. That is, we label the southeast boundary with the number $1, 2, \ldots, M$ starting at the southwest corner (i.e. the numbering is in the counterclockwise direction) so that the *right* pivots appear at the vertical border of the southeast boundary (see also Sect. 7.1.1). We construct the pipedream for $L^+(A)$ in the same way as that for the original $\mathcal{J}$-diagram (see Fig. 8.3). The set of trivalent vertices appearing in the pipedreams is then given by

$$\{v_{[1,2,3]}^{\circ}, v_{[4,5,6]}^{\circ}, v_{[1,5,7]}^{\circ}, v_{[4,6,7]}^{\circ}, v_{[1,2,7]}^{\bullet}, v_{[4,5,7]}^{\bullet}, v_{[5,6,7]}^{\bullet}, v_{[4,6,8]}^{\bullet}\}.$$

Using Algorithm 8.1, we can obtain the soliton graph $\mathscr{C}_+(\mathscr{M}(A))$ as shown in the right panel of Fig. 8.3.

Note here that two different $\mathcal{J}$-diagrams in Figs. 8.2 and 8.3 from the same chord diagram (i.e. ignoring the arrows) correspond to different expression of the same point in $\mathrm{Gr}(N, M)_{\geq 0}$. The matrix A calculated from the original $\mathcal{J}$-diagram (using the network representation) is given by

Fig. 8.3 The time-dual Le-diagram $L^+(A)$ with the pipedream and the soliton graph $\mathscr{C}_t(\mathscr{M}(A))$ for $t = 10$. Confirm that each pipe in the pipedream is uniquely identified with a line-soliton in the soliton graph

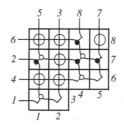

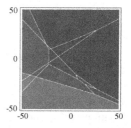

$$A = \begin{pmatrix} p_8 & p_5 & 1 & 0 & 0 & 0 & 0 & 0 \\ 0 & p_5 p_6 p_7 & p_6 p_7 & p_7 & 0 & -p_1 & -1 & 0 \\ 0 & 0 & 0 & p_3 p_4 & p_4 & 1 & 0 & 0 \\ 0 & 0 & 0 & 0 & 0 & p_1 p_2 & p_2 & 1 \end{pmatrix} \qquad \text{with}$$

1	1	p_8	1
p_7	1	p_6	p_5
1	p_4	p_3	
p_2	p_1		

For the new J-diagram $L^+(A)$ corresponding to the right chord diagram (the indices are in the reverse order), we can associate the pair (v', w') as

$$w' = s_2 s_1 s_5 s_4 s_3 s_2 s_6 s_5 s_4 s_3 s_7 s_6 s_5 s_4, \qquad v' = s_3 s_2 s_4 s_7 s_5 s_4.$$

Then from the pair (v', w'), one can obtain the matrix A' in the form,

$$A' = \begin{pmatrix} 1 & q_2 & q_1 q_2 & 0 & 0 & 0 & 0 & 0 \\ 0 & 0 & 0 & 1 & q_4 & q_3 q_4 & 0 & 0 \\ 0 & -1 & -q_1 & 0 & q_7 & (q_3+q_6)q_7 & q_5 q_6 q_7 & 0 \\ 0 & 0 & 0 & 0 & 0 & 1 & q_5 & q_8 \end{pmatrix} \qquad \text{with}$$

1	1	q_8	1
q_7	1	q_6	q_5
1	1	q_4	q_3
q_2	q_1		

Notice here that the matrix A' is in a *reversed* (or right) row echelon form and the *right* pivot set is $\{8, 7, 6, 3\} = w \cdot \{4, 3, 2, 1\}$. The matrix A' can be transformed from A to the form $A' = gA$ with some $g \in \mathrm{GL}_N$ and $\det(g) > 0$. Here, the parameters $(q_1, \ldots, q_8)$ are related to the p-parameters as

$$q_1 = \frac{1}{p_5}, \qquad q_2 = \frac{p_5}{p_8}, \qquad q_3 = \frac{1}{p_4}, \qquad q_4 = \frac{1}{p_3},$$

$$q_5 = \frac{1}{p_1}, \qquad q_6 = \frac{p_1 p_3}{p_7}, \qquad q_7 = \frac{1}{p_3 p_5 p_6}, \qquad q_8 = \frac{1}{p_1 p_2}.$$

Remark 8.2 For the KP hierarchy, the piecewise linear function is now given by

$$f_{\mathscr{M}(A)}(x, y, \mathbf{t}) = \max \{\Theta_I(x, y, \mathbf{t}) : I \in \mathscr{M}(A)\},$$

where $\Theta_I = \sum_{k=1}^{N} \theta_{i_k}$ for $I = \{i_1 < \cdots < i_N\}$ and

$$\theta_i(x, y, \mathbf{t}) = \kappa_i x + \kappa_i^2 y + \sum_{m=3}^{M-1} \kappa_i^m t_m \qquad \text{with} \quad t_3 = t.$$

Then it would be interesting to classify $\mathscr{C}_{\mathbf{t}}(\mathscr{M}(A))$ for each $\mathbf{t} \in \mathbb{R}^{M-3}$. Here we mention a result for $A \in \mathrm{Gr}(1, M)_{\geq 0}$ where the soliton graphs can be parametrized by the triangulations of M-gon and the soliton graph corresponding to each triangulation consists of the same set of line-solitons (see Problem 8.4 for the case $M = 5$ and Problem 8.4. Also see [57] for further details).

Problems

8.1 Find the soliton graphs $\mathscr{C}_\pm(\mathscr{M}(A))$ for all the cases for $\mathrm{Gr}(2, 4)_{\geq 0}$.

8.2 Let A be a 3×6 matrix from a cell associated with the derangement $\pi = (5, 1, 4, 6, 3, 2)$. Find the soliton graphs $\mathscr{C}_\pm(\mathscr{M}(A))$.

8.3 Consider the following graphs:

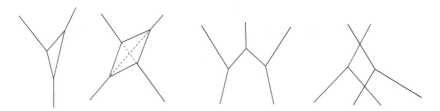

Show that the first two graphs cannot be soliton graphs. For the last two graphs, show that those can be soliton graphs and find the corresponding chord diagrams.

8.4 Consider a set of five points $\{\hat{\mathbf{v}}_i : i = 1, \ldots, 5\}$ on a parabola where each point $\hat{\mathbf{v}}_i$ is given by

$$\hat{\mathbf{v}}_i = (\kappa_i, \kappa_i^2, \omega_i(\mathbf{t})) \qquad \text{with} \qquad \omega_i(\mathbf{t}) = \kappa_i^3 t_3 + \kappa_i^4 t_4.$$

(a) Show the following formula for the index set $I = \{i_1 < \cdots < i_4\}$,

$$D_I(\mathbf{t}) := \left[(\hat{\mathbf{v}}_{i_2} - \hat{\mathbf{v}}_{i_1}) \times (\hat{\mathbf{v}}_{i_3} - \hat{\mathbf{v}}_{i_1})\right] \cdot (\hat{\mathbf{v}}_{i_4} - \hat{\mathbf{v}}_{i_1}) = K_I (t_3 + h_1(I)t_4),$$

where $K_I = \prod_{k>l}(\kappa_{i_k} - \kappa_{i_l})$ and $h_1(I) = \kappa_{i_1} + \cdots + \kappa_{i_4}$. (See Problem 3.3.)

(b) Following the argument on *regular* triangulation given in Problem 3.3, find the triangulations of the pentagon with vertices $\{v_i = (\kappa_i, \kappa_i^2) : i = 1, \ldots, 5)\}$. That is, construct the figure below, where we take $(\kappa_1, \ldots, \kappa_5) = (-2, -1, 0, 1, 2)$.

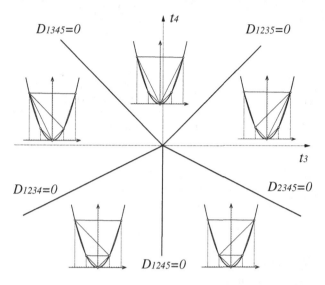

Each polyhedral region in the $t_3 t_4$-plane is given by the intersection of two regions $D_I \neq 0$ and $D_J \neq 0$ for certain index sets $\{I, J\}$. For example, the region including the points $(t_3 > 0, t_4 = 0)$ is given by the intersection of the regions with $D_{1235} > 0$ and $D_{2345} > 0$.

(c) Using the duality given in Problem 3.3 (see also Remark 1.2), construct a soliton graph associated to each triangulation given in the figure.

8.5 Discuss the connection between triangulation of M-gon inscribed in the parabola and the KP soliton graph for $N = 1$ (note that $M = 5$ for Problem 8.3). Here you may need to consider the KP hierarchy, i.e. you have $\theta_j(x, y, \mathbf{t}) = \kappa_j x + \kappa_j^2 y + \omega_j(\mathbf{t})$ with $\omega_j = \sum_{n=3}^{M-1} \kappa_j^n t_n$ for $j = 1, \ldots, M$. (See [57] for a further discussion).

8.6 In Problem 3.4, we considered the case with $N = 1$ and $M = 4$ and obtained the soliton graphs for $t > 0$ and $t < 0$. These graphs were constructed by the triangulations of 4-gon inscribed in the parabola. Now consider the case with $N = 2$ and $M = 4$ where the four points are given by $\hat{v}_i(t) = (v_i, \kappa_i^3 t) = (\kappa_i, \kappa_i^2, \kappa_i^3 t)$ for $i = 1, \ldots, 4$.

(a) Show that the convex hull of six points given by $\{\hat{v}_{ij} = \hat{v}_i(t) + \hat{v}_j(t) : 1 \leq i < j \leq 4\}$ is an *octahedron* in $\mathbb{R}^3$ if $t \neq 0$.

(b) Give the (regular) triangulation of the point set $\{v_{ij} = v_i + v_j : 1 \leq i < j \leq 4\}$, whose convex hull is a parallelogram by projecting the upper faces of the octahedron in (a) for $t \neq 0$.

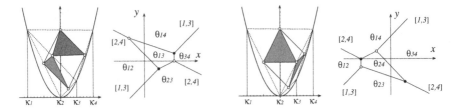

Fig. 8.4 Triangulations and the soliton graphs. The *left* two figures show the triangulation of the point set $\frac{1}{2}\{v_{12}, v_{23}, v_{34}, v_{14}, v_{13}, v_{24}\}$ (each point is shown as an open circle) and the corresponding soliton graph for $t < 0$. The *right* two figures are for $t > 0$. The κ-parameters are $(-2, 0, 1, 2)$. Each θ_i shows the dominant plane in that region. Note that the each *white* (*black*) *triangle* is dual to the *white* (*black*) vertex in the soliton graph (see also Fig. 3.5)

(c) Construct the soliton graphs for $t > 0$ and $t < 0$, which are generated by a totally positive 2×4 matrix A, i.e. the matroid is $\mathcal{M}(A) = \binom{[4]}{2}$. These soliton graphs are dual to the triangulations in (b) (see Fig. 8.4).

8.7 Let $\pi = (4, 5, 1, 2, 6, 3)$ be a derangement of S_6. Take the κ-parameter as $(\kappa_1, \ldots, \kappa_6) = (-1, -0.5, 0, 0.5, 1, 1.5)$. Then using Algorithm 8.1, construct the soliton graphs $\mathscr{C}_{\pm}(\mathcal{M}(A))$. Note that they correspond to the contour plots of the KP soliton shown in the preface.

References

1. Ablowitz, M.J., Segur, H.: Solitons and the Inverse Scattering Transform: SIAM Studies in Applied Mathematics. SIAM, Philadelphia (1981)
2. Ablowitz, M.J., Clarkson, P.A.: Solitons. Nonlinear Evolution Equations and Inverse Scattering. Cambridge University Press, Cambridge (1991)
3. Anker, D., Freeman, N.C.: On the soliton solutions of the Davey-Stewartson system for long wave. Proc. R. Soc. Lond. A **360**, 529–540 (1978)
4. Al-Salam, W.A., Ismail, M.E.H.: Orthogonal polynomials associated with the Rogers-Ramanujan continued fraction. Pacific J. Math. **104**, 269–283 (1983)
5. Adler, M., Cafasso, M., van Moerbeke, P.: Nonlinear PDEs for gap probabilities in random matrices and KP theory. Physica D **241**, 2265–2284 (2012)
6. Adler, M., van Moerbeke, P.: The spectrum of coupled random matrices. Ann. Math. **149**, 921–976 (1999)
7. Adler, M., van Moerbeke, P.: Vertex operator solutions to the discrete KP hierarchy. Commun. Math. Phys. **203**, 185–210 (1999)
8. Adler, M., van Moerbeke, P.: Generalized orthogonal polynomials, discrete KP and Riemann-Hilbert problem. Commun. Math. Phys. **207**, 589–620 (1999)
9. Adler, M., Shiota, T., van Moerbeke, P.: A Lax representation for the vertex operator and the central extension. Commun. Math. Phys. **171**, 547–588 (1995)
10. Airault, H., McKean, H.P., Moser, J.: Rational and elliptic solutions of the Korteweg-de Vries equation and a related many-body problem. Commun. Pure Appl. Math. **30**, 95–148 (1977)
11. Andrews, G.: Theory of Partitions. Encyclopedia of Mathematics & its Applications, Vol. 2. Addison-Wesley (1976)
12. Bertola, M., Eynard, B., Harnad, J.: Semiclassical orthogonal polynomials, matrix models and isomonodromic tau functions. Commun. Math. Phys. **263**, 401–437 (2006)
13. Berenstein, A., Fomin, S., Zelevinsky, A.: Parametrizations of canonical bases and totally positive matrices. Adv. Math. **122**, 49–149 (1996)
14. Balogh, F., Fonseca, T., Harnad, J.: Finite dimensional Kadomtsev-Petviashvili τ-functions. I. Finite Grassmannians. J. Math. Phys. **55** (2014) 083517 (32 pp)
15. Bjorner, A., Brenti, F.: Combinatorics of Coxeter groups, Graduate Texts in Mathematics, 231. Springer, New York (2005)
16. Biondini, G., Chakravarty, S.: Soliton solutions of the Kadomtsev-Petviashvili II equation. J Math. Phys. **47** (2006) 033514 (26 pp)
17. Biondini, G., Kodama, Y.: On a family of solutions of the Kadomtsev-Petviashvili equation which also satisfy the Toda lattice hierarchy. J. Phys. A: Math. Gen. **36**, 10519–10536 (2003)
18. Bóna, M.: Combinatorics of Permutations. Chapman & Hall/CRC, Boca Raton (2004)
19. Boiti, M., Pempinelli, F., Pogrebkov, A., Prinari, B.: Towards an inverse scattering theory for non decaying potentials of the heat equation. Inverse Probl. **17**, 937–957 (2001)

© The Author(s) 2017
Y. Kodama, *KP Solitons and the Grassmannians*,
SpringerBriefs in Mathematical Physics 22, DOI 10.1007/978-981-10-4094-8

20. Boiti, M., Pempinelli, F., Pogrebkov, A., Prinari, B.: The equivalence of different approaches for generating multisoliton solutions of the KPII equation. Theor. Math. Phys. **165**, 1237–1255 (2010)

21. Calogero, F.: Solution of the one-dimensional N-body problems with quadratic and/or inversely quadratic pair potentials. J. Math. Phys. **12**, 419–436 (1971)

22. Calogero, F., Degasperis, A.: Solitons and the Spectral Transform 1. North-Holland, Amsterdam (1982)

23. Casian, L., Kodama, Y.: Toda lattice, cohomology of compact Lie groups and finite Chevalley groups. Invent. Math. **165**, 163–208 (2006)

24. Casian, L., Kodama, Y.: On the cohomology of real Grassmann manifolds. arXiv:1309.5520

25. Chakravarty, S., Kodama, Y.: Classification of the line-solitons of KPII. J. Phys. A: Math. Theor. **41** (2008) 275209 (33 pp)

26. Chakravarty, S., Kodama, Y.: A generating function for the N-soliton solutions of the Kadomtsev-Petviashvili II equation. Contemp. Math. **471**, 47–67 (2008)

27. Chakravarty, S., Kodama, Y.: Soliton solutions of the KP equation and applications to shallow water waves. Stud. Appl. Math. **123**, 83–151 (2009)

28. Chakravarty, S., Kodama, Y.: Line-soliton solutions of the KP equation. AIP Conf. Proc. **1212**, 312–341 (2010)

29. Chakravarty, S., Kodama, Y.: Construction of KP solitons from wave patterns. J. Phys. A: Math. Theor. **47** (2014) 025201 (17 pp)

30. Chakravarty, S., Lewkow, T., Maruno, K.-I.: On the construction of KP line-solitons and their interactions. Appl. Anal. **89**, 529–545 (2010)

31. Corteel, S.: Crossings and alignments of permutations. Adv. Appl. Math. **38**, 149–163 (2007)

32. De Loera, J.A., Rambau, J., Santos, F.: Triangulations. Algorithm and Computation in Mathematics. Springer, Berlin (2010)

33. Deodhar, V.: On some geometric aspects of Bruhat orderings I. A finer decomposition of Bruhat cells. Invent. Math. **79**(3), 499–511 (1985)

34. Deodhar, V.: On some geometric aspects of Bruhat orderings II. The parabolic analogue of Kazhdan-Lusztig polynomials. J. Algebra **111**, 483–506 (1987)

35. Date, E., Jimbo, M., Kashiwara, M., Miwa, T.: Transformation groups for soliton equations in Nonlinear integrable systems–classical theory and quantum theory, pp. 39–119. World Scientific, Kyoto, Singapore (1981)

36. Davey, A., Stewartson, K.: On three-dimensional packets of surface waves. Proc. R. Soc. Lond. **A338**, 101–110 (1974)

37. Dickey, L.A.: Soliton Equations And Hamiltonian Systems. Advanced Series in Mathematical Physics, vol. 12. World Scientific, Singapore (1991)

38. Dimakis, A., Müller-Hoissen, F.: KP line-solitons and Tamari lattices. J. Phys. A: Math. Theor. **44**, 025203 (2011) (49 pp)

39. Enolski, V., Harnad, J.: Schur function expansions of KP tau functions associated to algebraic curves. Russian Math. Surv. **66**, 767–807 (2011)

40. Errera, A.: Une problème d'énumeration, Mém. Acad. Roy. Belgique Coll. 80, (2) **38** (1931) (26 pp)

41. Flajolet, P.: Analytic Combinatorics of Chord Diagrams. In: Formal Power Series and Algebraic Combinatorics (Moscow 2000), pp. 191–201. Springer, Berlin/New York (2000)

42. Freeman, N.C., Gilson, C.R., Nimmo, J.J.C.: Two-component KP hierarchy and the classical Boussinesq equation. J. Phys. A: Math. Gen. **23**, 4794–4803 (1990)

43. Freeman, N.C., Nimmo, J.J.C.: Soliton-solutions of the Korteweg-deVries and Kadomtsev-Petviashvili equations: the Wronskian technique. Phys. Lett. A **95**, 1–3 (1983)

44. Fulton, W.: Young Tableaux, London Mathematical Society Student Texts, vol. 35. Cambridge University Press (1997)

45. Fulton, W., Harris, J.: Representation Theory—A First Course. Springer (1991)

46. Fukuma, M., Kawai, H., Nakayama, R.: Infinite-dimensional Grassmannian structure of two-dimensional quantum gravity. Commun. Math. Phys. **143**, 371–403 (1992)

47. Fomin, S., Zelevinsky, A.: Double Bruhat cells and total positivity. J. Am. Math. Soc. **12**, 335–380 (1999)
48. Gardner, C.S., Greene, J.M., Kruskal, M.D., Miura, R.M.: Method for solving the Korteweg-de Vries equation. Phys. Rev. Lett. **19**, 1095–1097 (1967)
49. Griffiths, P., Harris, J.: Principles of Algebraic Geometry. Chelsea, New York (1978)
50. Goulden, I.P., Jackson, D.M.: The KP hierarchy, branched covers, and triangulations. Adv. Math. **219**, 932–951 (2008)
51. Gekhtman, M., Kasman, A.: On KP generators and the geometry of HBDE. J. Geom. Phys. **56**, 282–309 (2006)
52. Guil, F., Manas, M.: Finite-rank constraints on linear flows and the Davey-Stewartson equation. J. Phys. A: Math. Gen. **28**, 1713–1726 (1995)
53. Harada, H.: New subhierarchies of the KP hierarchy in the Sato theory. 2. Truncation of the KP hierarchy. J. Phys. Soc. Jpn **56**, 3847–3852 (1987)
54. Harnad, J., Orlov, A.: Scalar products of symmetric functions and matrix integrals. Theor. Math. Phys. **137**, 1676–1690 (2003)
55. Hirota, R.: Discrete analogue of a generalized Toda equation. J. Phys. Soc. Jpn **50**, 3785–3791 (1981)
56. Hirota, R.: The Direct Method in Soliton Theory. Cambridge University Press, Cambridge (2004)
57. Huang, J.: Classification of soliton graphs on totally positive Grassmannian. PhD thesis, Department of Mathematics, The Ohio State University (2015)
58. Ismail, M.E.H.: Classical and Quantum Orthogonal Polynomials on One Variable. Cambridge University Press, Cambridge, UK (2005)
59. Ismail, M.E.H., Stanton, D., Viennot, G.: The combinatorics of q-Hermite polynomials and the Askey-Wilson integral. Eur. J. Combin. **8**, 379–392 (1987)
60. Jimbo, M., Miwa, T.: Solitons and infinite dimensional algebra, vol. 19, pp. 943–1001. Publ. RIMS, Kyoto University (1983)
61. Kadomtsev, B.B., Petviashvili, V.I.: On the stability of solitary waves in weakly dispersive media. Sov. Phys. -Dokl. **15**, 539–541 (1970)
62. Kao, C-.Y., Kodama, Y.: Numerical study of the KP equation for non-periodic waves Math. Comput. Simul. **82**, 1185–1218 (2012)
63. Kasman, A.: Glimpses of Soliton Theory: The Algebra and Geometry of Nonlinear PDEs. Student Mathematical Library, AMS (2010)
64. Kac, V.: Infinite Dimensional Lie Algebras. Cambridge University Press (1990)
65. Kasraoui, A., Zeng, J.: Distribution of crossings, nestings and alignments of two edges in matchings and partitions. Elect. J. Comb. **13**, 1–12 (2006)
66. Kazhdan, D., Lusztig, G.: Schubert varieties and Poincaré duality. Geometry of the Laplace operator (Proc. Sympos. Pure Math., Univ. Hawaii, Honolulu, Hawaii, 1979). In: Proceedings of the Symposium on Pure Mathematics, vol. XXXVI, pp. 185–203. American Mathematical Society, Providence, RI (1980)
67. Kazarian, M.: KP hierarchy for Hodge integrals. Adv. Math. **221**, 1–21 (2009)
68. Kim, J., Lee, S.: Positroid stratification of orthogonal Grassmannian and ABJM amplitudes. J. High Energy Phys. (2014). doi:10.1007/JHEP09085
69. Kodama, Y.: Young diagrams and N-soliton solutions of the KP equation. J. Phys. A: Math. Gen. **37**, 11169–11190 (2004)
70. Kodama, Y.: KP solitons in shallow water. J. Phys. A: Math. Theory **43**, 434004 (2010) (54 pp)
71. Kodama, Y., Williams, L.: KP solitons, total positivity, and cluster algebras. Proc. Natl. Acad. Sci. USA **108**(22), 8984–8989 (2011)
72. Kodama, Y., Williams, L.: The Deodhar decomposition for the Grassmannian and the regularity of KP solitons. Adv. Math. **244**, 979–1032 (2013)
73. Kodama, Y., Williams, L.: KP solitons and total positivity for the Grassmannian. Invent. Math. **198**, 637–699 (2014)
74. Kodama, Y., Yeh, H.: The KP theory and Mach reflection. J. Fluid Mech. **800**, 766–786 (2016)

75. Konopelchenko, B.G.: Solitons in Multidimensions: Inverse Spectral Transform Method. World Scientific (1993)

76. Krichever, I.M.: Methods of algebraic geometry in the theory of non-linear equations. Russian Math. Surv. **32**, 185–213 (1977)

77. Krichever, I.M.: Rational solutions of the Kadomtsev-Petviashvili equation and integrable systems of N particles on a line. Funct. Anal. Appl. **12**, 59–61 (1978)

78. Lax, P.D.: Integrals of nonlinear equations of evolution and solitary waves. Commun. Pure Appl. Math. **21**, 467–490 (1968)

79. Li, W., Yeh, H., Kodama, Y.: On the Mach reflection of a solitary wave: revisited. J. Fluid. Mech. **672**, 326–357 (2011)

80. Lusztig, G.: Total positivity in reductive groups. In: Lie Theory and Geometry. Honor of Bertram Kostant, Progress in Mathematics, vol. 123. Birkhauser (1994)

81. Lusztig, G.: Total positivity in partial flag manifolds. Represent. Theory **2**, 70–78 (1998)

82. Macdonald, I.G.: Symmteric Functions and Hall Polynomials, Oxford Mathematical Monographs, 2nd edn. Oxford Science Publications (1995)

83. Marsh, R., Rietsch, K.: Parametrizations of flag varieties. Represent. Theory **8**, 212–242 (2004)

84. Matveev, V.B.: Darboux transformation and explicit solutions of the Kadomtsev-Petviaschvily equation. Lett. Math. Phys. **3**, 213–216 (1979)

85. Matveev, V.B., Salle, M.A.: Darboux Transformations and Solitons. Springer, Berlin (1991)

86. Medina, E.: An N soliton resonance for the KP equation: interaction with change of form and velocity. Lett. Math. Phys. **62**, 91–99 (2002)

87. Miles, J.W.: Obliquely interacting solitary waves. J. Fluid Mech. **79**, 157–169 (1977)

88. Miles, J.W.: Resonantly interacting solitary waves. J. Fluid Mech. **79**, 171–179 (1977)

89. Miura, R.: The Korteweg-de Vries equation, a survey of results. SIAM Rev. **18**, 412–459 (1976)

90. Miwa, T.: On Hirota's difference equation. Proc. Jpn. Acad. **58** Ser. A, 9–12 (1982)

91. Miwa, T., Jimbo, M., Date, E.: Solitons: differential equations, symmetries and infinite-dimensional algebras. Cambridge University Press, Cambridge (2000)

92. Mizumachi, T.: Stability of line solitons for the KP-II equation in $\mathbb{R}^2$. Memoirs of the American Mathematical Society, AMS (2015)

93. Molinet, L., Saut, J.C., Tzvetkov, N.: Global well-posedness for the KP-II equation on the background of a non-localized solution. Ann. Inst. H. Poincaré Anal. Non Lineéaire **28**, 653–676 (2011)

94. Moser, J.: Three integrable Hamiltonian systems connected with isospectral deformations. Adv. Math. **16**, 197–220 (1975)

95. Mulase, M.: Cohomological structure in soliton equations and Jacobian varieties. J. Diff. Geom. **19**, 403–430 (1984)

96. Mulase, M.: Matrix integrals and integrable systems. In: Fukaya, K. et. al. (eds.) Topology, Geometry and Field Theory, pp. 111–127. World Scientific (1994)

97. Newell, A.C.: Solitons in mathematics and physics. In: CBMS-NSF Regional Conference Series in Applied Mathematics, vol. 48. SIAM, Philadelphia

98. Novikov, S., Manakov, S.V., Pitaevskii, L.P., Zakharov, V.E.: Theory of Solitons: The Inverse Scattering Method. Contemporary Soviet Mathematics, Consultants Bureau, New York, London (1984)

99. Nishinari, K., Satsuma, J.: A new-type of soliton solution behavior in a two dimensional plasma system. J. Phys. Soc. Jpn. **62**, 2021–2029 (1993)

100. Ohta, Y., Satsuma, J., Takahashi, D., Tokihiro, T.: An elementary introduction to Sato theory. Prog. Theor. Phys. Suppl. **94**, 210–241 (1988)

101. Ohta, Y., Yang, J.: Dynamics of rogue waves in the Davey-Stewartson II equation. J. Phys. A. Math. Theor. **46**, 105202 (2013) (19 pp)

102. Penaud, J.-G.: Une preuve bijective d'une formule de Touchard-Riordan. Discrete Math. **139**, 347–360 (1995)

103. Postnikov, A.: Total Positivity, Grassmannians, and Networks. arXiv:math.CO/0609764

104. Ramanujan, S.: Notebooks, (2 volumes). Tata Institute of Fundamental Reasearch, Bombay (1957)
105. Rietsch, K.: Total positivity and real flag varieties. Ph.D. Dissertation, MIT (1998)
106. Riordan, J.: The distribution of crossings of chords joining pairs of $2n$ points on a circle. Math. Comput. **29**, 215–222 (1975)
107. Rogers, L.J.: Second memoir on the expression of certain infinite products. Proc. Lond. Math. Soc. **25**, 318–343 (1894)
108. Roselle, D.P.: Permutations by number of rises and successions. Proc. Am. Math. Soc. **19**, 8–16 (1968)
109. Russell, J.S.: Report on waves. Rept. Fourteenth Meeting of the British Association for the Advancement of Science, pp. 311–390. John Murray, London (1844)
110. Russell, J.S.: The Modern System of Naval Architecture, vol. 1, p. 208. Day and Son, London (1865)
111. Sato, M.: Soliton equations as dynamical systems on an infinite dimensional Grassmannian manifold. RIMS Kokyuroku (Kyoto University) **439**, 30–46 (1981)
112. Sato, M.: Soliton equations and universal Grassmannian varieties. Lecture notes taken by M. Noumi (in Japanese), Saint Sophia Univ. Lecture Notes, vol. 18 (1984)
113. Sato, M.: Lecture notes by Mikio Sato, Lecture notes taken by N. Umeda (in Japanese), Kyoto Univ. RIMS Lecture Notes, vol. 5 (1989)
114. Sato, M., Sato, Y.: Soliton equations as dynamical on infinite dimensional Grassmann manifold. In: Lax, P.D., Fujita, H., Strang, G. (eds.) Nonlinear Partial Differential Equations in Applied Sciences, pp. 259–271. North-Holland, Amsterdam, Kinokuniya, Tokyo (1982)
115. Satsuma, J.: N-soliton solution of the two-dimensional Korteweg-de Vries equation. J. Phys. Soc. Jpn. **40**, 286–290 (1976)
116. Satsuma, J.: A Wronskian representation of N-soliton solutions of nonlinear evolution equations. J. Phys. Soc. Jpn. **46**, 356–360 (1979)
117. Shiota, T.: Characterization of Jacobian varieties in terms of soliton equations. Invent. Math. **83**, 333–382 (1986)
118. Shiota, T.: Calogero-Moser hierarchy and KP hierarchy. J. Math. Phys. **35**, 5844–5849 (1994)
119. Stembridge, J.: On the fully commutative elements of coxeter groups. J. Algebr. Comb. **5**, 353–385 (1996)
120. Stanley, R.P.: Enumerative Combinatorics, vol. 2. Cambridge Studies in Advanced Math., no. 62, Cambridge University Press, Cambridge, UK (1997)
121. Takaoka, H.: Global well-posedness for the Kadomtsev-Petviashvili II equation. Discrete Contin. Dyn. Syst. **6**, 483–499 (2000)
122. Takasaki, K.: Auxiliary linear problem, differential Fay identities and dispersionless limit of Pfaff-Toda hierarchy. SIGMA **5**, 109 (2009) (34 pp)
123. Takasaki, K.: Differential Fay identities and auxiliary linear problem of integrable hierarchies. Adv. Stud. Pure Math. **61**, 387–441 (2011)
124. Talaska, K.: Combinatorial formulas for Γ-coordinates in a totally nonnegative Grassmannian. J. Combin. Theory Ser. A **118**(1), 58–66 (2011)
125. Touchard, J.: Sur une problème de configurations et sur les fractions continues. Can. J. Math. **4**, 2–25 (1952)
126. Tokihiro, T., Takahashi, D., Matsukidaira, J., Satsuma, J.: From soliton equations to integrable cellular automata through a limiting procedure. Phys. Rev. Lett. **76**, 3247–3250 (1996)
127. Talaska, K., Williams, L.: Network parameterizations of Grassmannians. Alg. Num. Theory **6**, 2275–2311 (2013)
128. Ueno, K., Takasaki, K.: Toda lattice hierarchy. Adv. Stud. Pure Math. **4**, 1–94 (1984)
129. Vanhaecke, P.: Stratifications of hyperelliptic Jacobians and the Sato Grassmannian. Acta Appl. Math. **40**, 143–172 (1995)
130. Villarroel, J., Ablowitz, M.J.: On the initial value problem for the KPII equation with data that do not decay along a line. Nonlinearity **17**, 1843–1866 (2004)
131. Whitham, G.B.: Linear and nonlinear waves. Wiley-Interscience, New York (1974)

132. Willox, R., Satsuma, J.: Sato theory and transformation groups. A unified approach to integrable systems. Lect. Notes Phys. **644**, 17–55 (2004)

133. Williams, L.K.: Enumeration of totally positive Grassmann cells. Adv. Math. **190**, 319–342 (2005)

134. Witten, E.: Quantum field theory, Grassmannians, and algebraic curves. Commun. Math. Phys. **113**, 529600 (1988)

135. Yaglom, A.M., Yaglom, I.M.: Challenging mathematical problems with elementary solutions, vol. I: Combinatorial Analysis and Probability Theory, Holden-Day, San Francisco, CA (1964)

136. Zakhalov, V.E., Fadeev, L.D.: The Korteweg-de Vries equation: a completely integrable Hamiltonian system. Funct. Anal. Appl. **5**, 280–287 (1971)

137. Zabusky, N.J., Kruskal, M.D.: Interaction of "solitons" in a collisionless plasma and the recurrence of initial states. Phys. Rev. Lett. **15**, 240–243 (1965)

Printed in the United States
By Bookmasters